FRAGILE AND FAILING STATES

SANDHURST TRENDS IN INTERNATIONAL CONFLICT

Edited by David Brown, Donette Murray, Malte Riemann, Norma Rossi and Martin A. Smith, The Royal Military Academy Sandhurst, UK

The Sandhurst Trends in International Conflict series is a cutting-edge forum and platform for original thought and debate on military and security matters within the contemporary international security environment. It aims to stimulate authors to think critically about contemporary conflict and security more generally and to identify and evaluate practical, political and doctrinal lessons from recent experience. The Sandhurst series invites practitioners and academics from the fields of security, diplomacy, the law, politics and the military to interrogate and publish on the key debates that will shape both the contemporary international security environment and a modern military operating within it.

Fragile and Failing States
Challenges and Responses

Edited by
DAVID BROWN
DONETTE MURRAY
MALTE RIEMANN
NORMA ROSSI
MARTIN A. SMITH

Howgate Publishing Limited

First published in 2020 by
Howgate Publishing Limited
Station House
50 North Street
Havant
Hampshire
PO9 1QU
Email: info@howgatepublishing.com
Web: www.howgatepublishing.com

British Library Cataloguing-in-Publication Data
A catalogue record for this book is available from the British Library

ISBN 978-1-912440-19-1 (pbk)
ISBN 978-1-912440-16-0 (ebk - PDF)
ISBN 978-1-912440-17-7 (ebk - ePUB)

The views expressed in this book are those of the individual authors and do not necessarily reflect official policy or position.

CONTENTS

FIGURES AND TABLES

Figures

Tables

ABOUT THE EDITORS

Dr David Brown is a Senior Lecturer in the Department of Defence and International Affairs at the Royal Military Academy Sandhurst. He has written extensively on a range of security-related issues, publishing books and articles on US and UK foreign and defence policy, contemporary power relations, aspects of European security and international intervention.

Dr Donette Murray is a Senior Lecturer at the Royal Military Academy Sandhurst. A Fellow of the Supreme Headquarters Allied Powers Europe (SHAPE) and a former political advisor, she holds a doctorate from the University of Ulster and an LLM in International Law from the University of Maastricht. The author of four books on US foreign policy, she has also published in the *Hague Yearbook of International Law* on states' use of force in self-defence against non-state actors.

Dr Malte Riemann is a Senior Lecturer in the Department of Defence and International Affairs at the Royal Military Academy Sandhurst. He studied in Bremen and Pietermaritzburg and holds a PhD in International Relations from the University of Reading. His fields of interest include the privatisation of war and its effects on the state's legitimate monopoly on violence, the medicalisation of security, and the historicity of non-state actors.

Dr Norma Rossi is a Senior Lecturer in Defence and International Affairs at the Royal Military Academy Sandhurst. She studied in Rome and Paris and received her PhD in Politics and International Relations from the University of Reading on an Earhart Foundation fellowship. Her research interests include transnational organised crime, the role of far-right parties in security discourses, the changing character of conflict, and the role of Professional Military Education in Security Sector Reform.

Dr Martin A. Smith is Senior Lecturer in Defence and International Affairs at the Royal Military Academy Sandhurst. Prior to joining RMAS he was at the Department of Peace Studies University of Bradford, from where he received his PhD in 1994. His main research interests are in the fields of international power, European security and US foreign policy and he has published widely in these areas.

CONTRIBUTORS

David Brown is a Senior Lecturer in the Department of Defence and International Affairs at the Royal Military Academy Sandhurst. He has written extensively on a range of security-related issues, publishing books and articles on US and UK foreign and defence policy, contemporary power relations, aspects of European security and international intervention.

David Chandler is Professor of International Relations at the University of Westminster. He is the author of a number of books on international state-building, including *International Statebuilding: The Rise of Postliberal Governance* (2010) and *Empire in Denial: The Politics of Statebuilding* (2006) and is the co-editor (with Timothy Sisk) of the *Routledge Handbook of International Statebuilding* (2013).

Jonathan Fisher is Reader in African Politics in the International Development Department at the University of Birmingham. He is also a Research Fellow in the Centre for Gender and Africa Studies at the University of the Free State. His work focuses on the intersection between authoritarianism, conflict and insecurity and he has a particular interest in East Africa, having conducted research in Djibouti, Eritrea, Ethiopia, Kenya, Rwanda, Somaliland and Uganda. He is the author of 13 articles on these themes and of two forthcoming books – *East Africa after Liberation: Conflict, Security and the State since the 1980s* and (with Nic Cheeseman) *Authoritarian Africa: Repression, Resistance and the Power of Ideas.*

Islam Mohamed Goher is a graduate student at Cairo University and City University New York. In 2006, he obtained his BSc degree in Political Science from the Faculty of Economics and Political Science at Cairo University. In

2011, he obtained his first Master's degree, in political science with distinction, from the same faculty. His dissertation was on 'Continuity and Change in United States' Foreign Policy after September 11, 2001'. In 2013, he joined the PhD programme at Cairo University, where he has been developing a model on the relationship between foreign intervention and the process of state failure and is expected to complete his thesis in 2020. In 2019, he was awarded his second Master's degree, in International Relations, with distinction, from City University New York. In his dissertation he applied the relational model of foreign intervention and the process of state failure, which he has been developing in his PhD, to the foreign intervention in Libya in 2011.

Emily Knowles is the Director of the Oxford Research Group's Remote Warfare Programme. She writes and speaks regularly on changes in military engagement, and specifically the shift towards states like the UK playing an increasingly supporting role alongside local and regional forces who do the bulk of frontline fighting. She also leads on the team's field research, having recently conducted research in Bamako, Nairobi, Kabul, Basra, and Baghdad. Her commentary has been included in programmes like BBC1's *The Big Questions*, BBC Radio 4's *The Briefing Room*, and the British Forces Broadcasting Services' *Sitrep*.

Lt Col Jahara 'Franky' Matisek completed his PhD in Political Science in 2018 at Northwestern University, focusing on how weak states create effective armies in Africa. He is an Assistant Professor of Military and Strategic Studies at the US Air Force Academy and is a two-time (2018–19 & 2019–20) Non-Resident Fellow with the Modern War Institute at West Point, US Military Academy. He is a mobility pilot with over 3,000 hours of flight time, including over 200 combat missions in Afghanistan and Iraq, and was a deployed planner supporting the 2011 Libyan No Fly Zone. In 2019, Lt Col Matisek will become the Director of the Center for Airpower Studies (CAPS) at the US Air Force Academy and will deploy to Afghanistan as a pilot on the E-11 Battlefield Airborne Communications Node (BACN). Lt Col Matisek has published over 40 articles on warfare and strategy in peer-reviewed journals and policy relevant outlets.

Jan Pospisil is Head of Research at the Austrian Study Centre for Peace and Conflict Resolution (ASPR) in Vienna. He is a partner of the Political Settlements Research Programme (PSRP) at the University of Edinburgh and involved in the development of the PA-X peace agreements database.

His research focuses on peace processes, political settlements, humanitarian negotiations, resilience, and Sudanese and South Sudanese politics. He focuses in particular on the consequences of complexity thinking on peace-building concepts. His monograph, *Peace in Political Unsettlement*, has recently been published.

Saúl M. Rodríguez is a PhD Fellow in the University of Ottawa´s Political Studies School. He has been researcher and professor in several universities in Colombia. He has published widely in Spanish and English on militarism, democracy, war studies, peace and conflict, international relations, and counterculture. Currently he is conducting research into militarism and democracy in the Global South. He holds a BA in History from the National University of Colombia and an MA in Social Sciences investigation from the University of Buenos Aires.

Fabio Sánchez is Professor and Research Head at the School of Politics and International Relations, Sergio Arboleda University in Colombia. His research work concentrates on Colombian foreign policy, multilateralism and regional integration in South America. His most recent published book is *Unasur: poder y acción en Suramérica*. He is interested in the relationship between Colombian foreign policy and the internal armed conflict. He holds a PhD in International Relations from the Autonomous University of Barcelona.

Jacob Thomas-Llewellyn attended Reading University between 2012–17, attaining a First Class Honours Degree in War, Peace and International Relations and winning the Peter Campbell Prize for Writing Excellence. He was subsequently awarded a research grant by the Department of Politics and International Relations to study at Master´s level which he completed with a distinction in Strategic Studies, winning the Department's award for best dissertation. The basis of his dissertation was an analysis of the Normandy landings to determine the influence of logistics on the decision making and planning for Operation OVERLORD. Now undertaking his doctoral research in history, his thesis is examining British political, military and industrial relationships during the Second World War, using Projects MULBERRY and PLUTO as case studies. He is being sponsored by industry through the Wates Group and Sir Robert McAlpine Foundation Research Grants, allowing part of the study to be completed overseas, and the D-Day Story Research Grant for study at the D-Day Museum in Portsmouth. Since

2017 he has been attached to the Centre for Army Leadership at the Royal Military Academy Sandhurst as a Resident Researcher.

Abigail Watson is a Senior Research Officer at the Oxford Research Group's Remote Warfare Programme. She researches, writes and presents on the military, legal, and political implications of remote warfare. Her articles have been published in *Just Security, Strategy Bridge, E-IR, OpenDemocracy,* and the *Small Wars Journal.*

FOREWORD

This is an important book for any soldier, sailor or airman who thinks deeply about their profession.

The people of Britain's armed forces have long had to face multiple challenges, from preparing to defend our homeland to conducting small and irregular wars overseas. Recently, most of our conflicts have been in or around failed, failing or fragile states.

It is not inevitable that *all* future conflicts involving British armed forces will be in or around failed states, but there are three reasons why *most* of them will be. First, failed states can, like pre-9/11 Afghanistan or present-day Somalia, provide safe havens for terrorist organisations that export their violence to other states and can pose a direct threat to the UK or our friends. Second, failed and failing states are prone to humanitarian crises with which they are unable to cope and the international community may be forced, under the doctrine of responsibility to protect, to intervene. Third, the impact of state failure may encourage the mass migration of people which, in addition to humanitarian concerns, may jeopardise the stability of neighbouring and regional states.

For the first time since the end of the Cold War, we are also faced with the possibility of state-on-state warfare with peer competitors. But the reality of modern warfare is that such armed competition may itself play out in the environment of a failed state, using proxies or other deniable forms of warfare. So, a fourth reason why we may need to engage in a failing state is to protect a friend within the wider context of a new contest between great powers.

Those responsible for planning and leading operations in failed, failing or fragile states will gain much from this book: recognising what state failure is and what causes it; understanding the security challenges posed by state failure; examining the comparative benefits of who is best to intervene; and suggesting a range of intervention options. The book lays out the theory and then illustrates it through a variety of case studies.

As various contributors to the book point out, those responsible for deciding upon and then implementing a military intervention will certainly not understand the whole range of issues confronting a failed or failing state. As a result, and as Professor Sir Michael Howard pointed out, it is inevitable that mistakes will be made early in any campaign within that state. But, by understanding what those mistakes might be, we can hope to be near enough not to derail the campaign as a whole and adaptable enough to change our structures, systems and approach, as we gain the understanding and insight needed to turn the campaign into strategic success. That is the real challenge facing today's military professionals, as it was for yesterday's and, no doubt, it will be for tomorrow's.

Colonel Richard Iron CMG OBE

ACKNOWLEDGEMENTS

Edited books are, by definition, team efforts, and we are pleased to acknowledge with gratitude the commitment, hard work and patience of our contributors in seeing this project through to its successful completion. We are especially grateful for their diligence in submitting drafts and responding to editorial queries on, or even ahead of, deadline – a level of efficiency rarely encountered in a collective writing project!

Two people at RMAS deserve particular mention for their stalwart support for our STIC initiative to date. Brigadier Bill Wright will shortly leave his post as Commander Sandhurst Group and his interest in and enthusiasm for our project will be sorely missed. We are fortunate that Lieutenant Colonel Jane Hunter will be in her job for a while yet. Since the start, her support has been invaluable, if not indispensable, not least in helping us to publicise our symposia and forthcoming books through and within military networks. She has also been an endless and level-headed source of sound practical advice.

DB
DM
MR
NR
MAS
August 2019

ABBREVIATIONS

AMISOM	African Union Mission in Somalia
ANA	Afghan National Army
ANDSF	Afghan National Defence and Security Forces
ANP	Afghan National Police
APC	Armoured Personnel Carriers
AQAP	Al-Qaeda in the Arabian Peninsula
AU	African Union
AUC	Autodefensas Unidas de Colombia
CBL	Central Bank of Libya
CDS	Chief of Defence Staff
CIA	Central Intelligence Agency
CINC	Composite Index of National Capabilities
COW	Correlates of War project
CPA	Comprehensive Peace Agreement
DAC	Development Assistance Committee
DDR	disarmament, demobilisation and reintegration
DOE	Dispute Outcome Expectations
DRC	Democratic Republic of Congo
DS	Democratic Security
ELN	Ejército de Liberación Nacional (Colombia)
ENDF	Ethiopian National Defense Force
EPL	Ejército Popular de Liberación (Colombia)
EPRDF	Ethiopian People's Revolutionary Democratic Front
EU	European Union
EUTM-S	European Union Training Mission in Somalia
EVD	Ebola virus disease
FARC	Fuerzas Armadas Revolucionarias de Colombia

FATA	Federally Administered Tribal Areas
FBI	Federal Bureau of Investigation
FCO	Foreign and Commonwealth Office
FDI	foreign direct investment
FSI	Fragile States Index
GAO	Government Accountability Office
GDP	gross domestic product
GNA	Government of National Accord
GNC	General National Congress
GWOT	global war on terror
HOR	House of Representatives
HRC	Human Rights Council
IBL	institutionalisation before liberalisation
ICC	International Criminal Court
ICG	International Contact Group on Libya
ICU	Islamic Courts Union
IDP	internally displaced persons
IDPS	International Dialogue on Peacebuilding and Statebuilding
IGAD	Intergovernmental Authority on Development
IIAG	Ibrahim Index of African Governance
INCAF	International Network on Conflict and Fragility
IPI	International Peace Institute
IRC	International Red Cross
IS	Islamic State
ISIL	Islamic State of Iraq and the Levant
ISIS	Islamic State of Iraq and Syria
LIA	Libyan Investment Authority
LIFG	Libyan Islamic Fighting Group
LMB	Libyan Muslim Brotherhood
LNA	Libyan National Army
LPA	Libyan Political Agreement
LPTIC	Libyan Post, Telecommunication and Information Technology Company
MOAB	mother of all bombs
MSF	Médecins Sans Frontières
NATO	North Atlantic Treaty Organisation
NFA	National Forces Alliance
NGO	non-governmental organisation
NOC	National Oil Company

NRA	National Resistance Army
NSS	National Security Strategy
NTC	National Transitional Council
ODA	Overseas Development Assistance
ODI	Overseas Development Institute
OECD	Organisation for Economic Co-operation and Development
OUP	Operation Unified Protector
PA-X	Peace Agreement Access Tool
PC	Presidential Council
PMC	private military companies
PMF	Popular Mobilisation Forces
PSG	Peacebuilding and Statebuilding Goals
R2P	Responsibility to Protect
RDF	Rwanda Defence Force
RoE	rules of engagement
RPF	Rwandan Patriotic Front
SDC	Syrian Democratic Council
SDF	Syrian Democratic Forces
SF	Special Forces
SFA	security force assistance
SFG	Somali Federal Government
SIGAR	Special Inspector General for Afghanistan Reconstruction
SNA	Somali National Army
SSA	sub-Saharan Africa
SSR	security sector reform
TAA	Train, Advise and Assist
TFG	Transitional Federal Government
UAE	United Arab Emirates
UK	United Kingdom
UN	United Nations
UNAMIR	United Nations Assistance Mission for Rwanda
UNAP	United Nations Action Plan
UNHCR	United Nations High Commissioner for Human Rights
UNSC	United Nations Security Council
UNSCR	United Nations Security Council Resolution
UNSOM	United Nations Mission in Somalia
UPDF	Uganda People's Defence Force
US	United States

USSOF	United States Special Operations Forces
WHO	World Health Organization
YPG	Yekîneyên Parastina Gel (Syrian Kurds)

INTRODUCTION

Martin A. Smith[1]

The risks and challenges posed by 'fragile', 'failing' and 'failed' states have been enduring issues in the post-Cold War international security debates. For Western states, including the United Kingdom (UK), they have been an issue present virtually from the start. The collapse of central governmental control in Somalia during 1991 exacerbated a humanitarian crisis that initially drew a media-friendly response from the United States (US), in the form of food-aid deliveries, at the end of the George H.W. Bush administration's time in office in December 1992. From early 1993, therefore, the incoming Clinton administration inherited this commitment and indeed deepened it during the course of that year in conjunction with the United Nations (UN), but in a haphazard way that would culminate in the ill-thought-through crossing of what was later dubbed the 'Mogadishu line' between humanitarian relief and assistance, peacekeeping and peace-enforcement activities. This culminated in the debacle of the (in)famous 'Black Hawk down' incident in the Somali capital in October 1993. The scars left by this on the US political and military psyche have proved to be sufficiently deep and enduring as to impair American willingness to participate significantly in UN-led or sponsored peace operations over the course of nearly three decades and spanning four presidencies, although concerns about the impact of state fragility and failure more generally have certainly endured, and indeed were effectively given a new lease of life after the events of 11 September 2001 (9/11).

In key respects, therefore, Somalia proved to be the initial foundational experience, shaping and conditioning Western views about state fragility and collapse and their potential impacts. The Balkan wars that endured through much of the 1990s – in Bosnia from 1992–95 and Kosovo in

---

1   The views expressed here are personal and should not be taken to represent the policy or views of the British Government, Ministry of Defence or the Royal Military Academy Sandhurst.

1998–99 – were also instrumental, not least in shaping views and debates in the UK and among other European states. Europeans were barely involved in Somalia, but France and the UK were the two largest troop-contributing states to the UN relief and assistance mission in Bosnia during the mid-1990s, while several European member states were involved militarily in the North Atlantic Treaty Organisation's (NATO) coercive humanitarianism in Kosovo during March-June 1999. A broad cross-section of members also contributed to the long-term stabilisation forces that were deployed in both theatres following the conclusion of the conflicts there.

Outline and Plan of this Book

By the turn of the millennium, therefore, the importance of state fragility and failure, terms which may be broadly understood as denoting the withdrawal and – albeit more contentiously – withholding of the essential services that central state structures and authorities are expected to provide, was firmly established on the agenda of Western states, even before the events of 9/11 ushered in a 'second age' of concern about its potential consequences, as noted later. What could and should be done about it, however, was less clear and has remained so to the present day. The chapters that follow address key elements of the contemporary debates relevant to academics and practitioners. The first of these is essentially conceptual in the sense of critically reconsidering what state fragility and failure looks like, what the security consequences and implications of fragile and failing states are and how – and indeed whether – the consequences can be mitigated and addressed by outside intervention.

These issues are examined in Part One of the book. David Chandler opens with a challenge to those who think about these problems and those who take on the task of responding to them. He asks whether it is even possible, at least for outsiders, to understand in any meaningful and useful way what state failure is and what its essential causes are in any given context. As Chandler notes, this is a question with profound policy implications. If it is *not* possible for outsiders, by definition, to understand the phenomenon in any particular indigenous context, then this calls into question the basis for external interventions of all kinds, however well-intentioned.

In practice, the best that might be hoped for is to effectively freeze political and other forms of intra-state contestation within an open-ended peace process and, in doing so, enshrine, as Jan Pospisil suggests in

Chapter 2, a de facto 'political unsettlement', which at least has the virtue of setting in place a sustainable balance between competing groups and factions within a state and society. This can work, as seen in the relative success of the Dayton peace process in Bosnia since 1996. It is, however, inherently fragile – as witnessed by the unending debates over the future of Bosnia 25 years on from Dayton – and, in a number of prominent cases, such as Somalia and South Sudan, it has thus far proved impossible to institutionalise and sustain, at least on a territory-wide basis.

Another enduring – and apparently insoluble – problem can be seen in the unevenness and inconstancy of effort and attention devoted to the challenges of fragility and failure by other states and the international organisations and institutions that they may choose to operate through. This is surveyed here over several case studies by Jacob Thomas-Llewellyn in Chapter 3. In some respects, it represents the other side of the coin discussed by Chandler: if there is indeed no viable objective basis for external interventions, there are cases where it can be argued that the *absence* of intervention demonstrably makes things even worse. As Thomas-Llewellyn notes, this is illustrated perhaps most graphically by the consequences of international inaction over the unfolding Rwandan genocide in April–July 1994. Taken together therefore, the chapters in Part One raise and discuss key questions about the sources and consequences of both intervention *and* non-intervention in response to the fraying or collapse of effective and/or legitimate centralised state structures and processes.

Part Two consists of a series of case study-based analyses, each illustrating one or more of the important underlying issues in the contemporary state fragility debates. Chapters 4 and 5 both deal with the African context, sometimes assumed to be particularly afflicted by the challenges of state fragility, although in this context it is worth remembering that several of the key formulative international experiences of state failure during the 1990s happened in the Balkans in south-east Europe. Nevertheless, as Jahara Matisek notes in his chapter, by the measurements used by many analysts, African states do score relatively weakly. Both chapters in this mini-section amplify and develop the themes and issues raised initially in Part One. In Chapter 4, Jonathan Fisher, using the African Union (AU) mission in Somalia as his case study, builds on Chandler's earlier analysis by challenging the notion that 'local' intervenors are any better placed or more prepared than Westerners or other 'internationals' to understand, appreciate and thus respond effectively to the conditions breeding fragility and failure in particular states. In Fisher's estimation (as in that of Thomas-Llewellyn in the preceding chapter), a major issue is the extent to which intervening states per se act on the basis of their own perceived national interests, rather than through any manifestly genuine commitment to internationalist or humanitarian motivations. This is hardly a new observation, but it is a fundamental one in a world still ordered legally and structurally on the basis of individual sovereign states.

In Chapter 5, Matisek considers a number of African states in developing his core argument about the apparent dichotomy whereby 'weak' states nevertheless choose to develop and maintain 'strong', costly and well-equipped armed forces. His analysis uncovers a core distinction – later picked up and discussed by David Brown in the Conclusions – between states and *regimes*. Participants in the debates sometimes talk or write about 'states' generically when it is apparent that what they really have in mind are their governing regimes, which are by no means always cohesive or unitary actors. As Brown points out, it should not be assumed that centralised regime breakdown is necessarily coterminous with 'state failure' more generally. In Somalia, for example, effective state-like structures do exist and function on a local and regional basis – most obviously in the self-proclaimed Republic of Somaliland – even though there has been no legitimate or effective central state authority over the whole of Somalia's recognised sovereign territory for close to three decades.

The notion of uneven state authority and presence is further evaluated and developed in the context of contemporary Colombia by Saúl M. Rodríguez and Fabio Sánchez in Chapter 6. Their argument is based on the concept of 'failed state stigma', which is again reflective of tendencies to generalise when considering state weakness and inefficiency. In the view of Rodríguez and Sánchez, the centralised Colombian state has been *partially* effective in maintaining authority and a presence – in terms of basic service-provision – in some of the regions and areas under its sovereign jurisdiction, even during the worst years of insurgency by the *Fuerzas Armadas Revolucionarias de Colombia* (FARC) and other armed groups. Nevertheless, in their view, Colombia has often been portrayed by outsiders as a 'failed state' per se, which has impacted how other states and international institutions view and interact with it. Their argument also carries an echo from David Chandler's earlier analysis in that it touches on underlying notions of how ideas and concepts such as 'failed state' are constructed, by whom, for what purposes and with reference to what 'objective reality' – insofar as such a thing can be determined and measured.

In Chapter 7, Islam Goher provides a conceptual lens for analysing the international intervention in Libya in the context of the de facto civil war there during 2011. This was consequential for several reasons, not least because it was – and so far remains – the most significant intervention to be authorised by the UN partially with reference to the Responsibility to Protect (R2P) framework, which its members had formally adopted six years earlier. Secondly, the situation in Libya after 2011 offers another graphic example of the partial and selective nature of international interventions, witnessed in this case most clearly and consequentially by NATO's decision to declare 'mission accomplished' and leave as soon as Libyan leader Gaddafi was found and summarily executed by rebels in October 2011. As a result, as Goher argues, the intervention proved instrumental in inadvertently turning a weak Libyan state (masked by an apparently strong regime up to 2011) into a failed one,[2] with ensuing deleterious consequences for the Libyan people, regional security and wider stability in, for example, southern Europe, with an influx of migrants coming from and through violently contested Libyan territory.

As noted earlier, the 9/11 attacks in New York and Washington DC ushered in a 'second age' of Western debate about state fragility and failure and, especially, its consequences. Among the most prominent and contested elements of these newly-invigorated debates has been a focus on questions about whether, how and to what extent fragile and failing states provide a 'safe haven' for international terrorist groups like Al-Qaeda (AQ) and a base for consolidation for Islamic State (IS), and also how international (particularly Western) counterterrorism efforts can most effectively deal with these. In Chapter 8, Emily Knowles and Abigail Watson focus on these questions. Based on extensive fieldwork, including multiple interviews with military personnel on operations, their analysis develops a detailed critique of the effectiveness of the approach increasingly being adopted by the UK and its allies in, for example, Afghanistan and Syria, which they call 'remote warfare'. As such, their critique speaks directly to military and other practitioners engaged in responding to the ongoing challenges posed by the continuing international campaign against militant Islamist-inspired terrorist activity.

As Brown notes in the Conclusions, the debates on state fragility and failure over the past three decades have been enduring, extensive and

---

2   Given the particular nature and structure of the system of governance developed by Gaddafi over his 40 years in power, it does seem fair in this context to describe the Libyan 'state' as being essentially coterminous with his regime. The former had little, if any, effective autonomous existence beyond the latter, as evidenced by the depth and extent of the governance void that quickly became apparent after Gaddafi's capture and death.

sometimes contested, but, thus far, they have tended to generate much heat and rather less light. There is little if any agreement even on fundamental issues of how state fragility and failure should most effectively be measured, still less on what roles outside interveners – either regional or from the wider international community – can and should play in addressing the challenges they pose. In this context, if the contributions to this volume help to sharpen the focus of the debates, more effectively delineate some of their core components and provide impetus for further reflection and research, it will have served its purpose.

1

CHALLENGES TO INTERNATIONAL STATE-BUILDING

From Annihilation to Acceptance

David Chandler

Introduction

In the late 1990s and the 2000s, discourses of international state-building hailed a new policy framework: the so-called 'bottom-up' approach, which was to overcome the limits of overly prescriptive and generic international programmes that assumed that 'one size fits all'.[1] However, this alternative framework – one which presupposed difference as a starting point, rather than the uniformity of problems and solutions[2] – similarly has today reached an impasse. Alternative approaches seem mired in discussion of the problems

---

1   There is an extensive literature, summarised in Christina Bennett, Matthew Foley and Hanna B. Krebs, 'Learning from the past to shape the future: Lessons from the history of humanitarian action in Africa', *Humanitarian Policy Group Working Paper* (London: Overseas Development Institute, 2016); Sarah Collinson, 'Constructive deconstruction: making sense of the international humanitarian system', *Humanitarian Policy Group Working Paper* (London: Overseas Development Institute, 2016); David Chandler, *Peacebuilding: The Twenty Years' Crisis, 1997–2017* (Basingstoke: Palgrave Macmillan, 2017).
2   See Morgan Brigg, *The New Politics of Conflict Resolution: Responding to Difference* (Basingstoke: Palgrave Macmillan, 2008).

of 'relational sensitivity',[3] the 'local turn',[4] 'hybridity',[5] 'friction'[6] or new forms of representation and inclusion.[7] In fact, as long as these alternatives still shared the assumptions of external knowledge and direction, the attention to difference merely multiplied the points of impasse, bringing into focus further limits to external policy practices and forms of knowledge.[8]

The focus of this chapter is the analysis of a discursive shift away from these 'external' articulations of alternative 'bottom-up' understandings of international state-building and towards what are described here as exploratory or 'affirmative' forms of knowing and agency. This can be seen as a two-stage process. Firstly, with the opening up of the 'black box' of endogenous social processes,[9] there is the growing recognition of the limits of 'bottom-up' forms of state-building intervention, as an attempt to establish a new ground for external problem-solving. Secondly, there is a shift to 'the great outdoors' through imagining alternative ways of perceiving and responding: accepting problems as emergent and interactive processes that are invitations to grasp the world in richer and more complex ways that are much more affirmative of the limits of knowing and the diversity of contexts and situations.

The 'Bottom-Up' Approach

There is a growing policy convergence in international approaches to state-building intervention, increasingly covering the fields of peace and security,

---

3 Wren Chadwick, Tobias Debiel, and Frank Gadinger (eds), 'Relational Sensibility and the "Turn to the Local": Prospects for the Future of Peacebuilding', *Global Dialogues* 2 (Duisburg: Käte Hamburger Kolleg/Centre for Global Cooperation Research, 2013).
4 Elisa Randazzo, *Beyond Liberal Peacebuilding: A Critical Exploration of the Local Turn* (Abingdon: Routledge, 2017).
5 Gearoid Millar, 'Disaggregating hybridity: Why hybrid institutions do not produce predictable experiences of peace', *Journal of Peace Research* 51:4 (2015), 501-14; Suthaharan Nadarajah and David Rampton, 'The limits of hybridity and the crisis of liberal peace', *Review of International Studies* 41:1 (2015), 49–72.
6 Annika Björkdahl et al. (eds), *Peacebuilding and Friction: Global and Local Encounters in Post-Conflict Societies* (Abingdon: Routledge, 2016).
7 For example, Jenny H. Peterson, 'Creating Space for Emancipatory Human Security: Liberal Obstructions and the Potential of Agonism', *International Studies Quarterly* 57:2 (2012), 318–28.
8 See, for example, Tobias Debiel, Thomas Held and Ulrich Schneckener (eds), *Peacebuilding in Crisis: Rethinking paradigms and practices of transnational cooperation* (Abingdon: Routledge, 2016); Volker M. Heins, Kai Koddenbrock and Christine Unrau (eds), *Humanitarianism and Challenges of Cooperation* (Abingdon: Routledge, 2016).
9 A 'black box' is a complex system or device whose internal or endogenous processes are hidden from view or understanding.

development and environmental sustainability,[10] and cohered through the United Nations' (UN) 2030 Agenda for Sustainable Development.[11] In 2016, the then UN Secretary-General, Ban Ki-moon, argued that policy-makers need to move 'beyond short-term, supply-driven response efforts towards demand-driven outcomes'.[12] This trope of moving beyond 'supply-driven' responses problematises the established frameworks and institutional arrangements of international state-building, breaking down the silos of expertise and authoritative knowledge that have traditionally been key to legitimising international policy prescriptions. However, the difficulty of relegitimising and repurposing international state-building is unlikely to be resolved merely through inversing the focus and direction of agency from the 'top-down' external 'solution' to the 'bottom-up' consensual understanding of the local 'problem'.

As the Overseas Development Institute (ODI) has highlighted, the shift towards bottom-up approaches is driven by the perception that international state-building agencies face a crisis of legitimacy, one that goes to the heart of their identity and the belief international policy interventions can be neutral or objective in the desire to problem-solve and capacity-build, 'regardless of context or culture'.[13] This idea of Western ethics, expertise and knowledge as applicable universally was crucial to state-building and to liberal internationalist approaches to peace and development assistance. However, it is seen to be problematic today and to represent a 'Western ethos' that others would wish to 'question or reject'.[14]

> Large parts of the current way the West conducts its … business
> – the reluctance to properly engage with and respect local
> authorities and cultures, the tendency to privilege international
> technical expertise over local knowledge and capacities, with

---

10 See *The Millennium Development Goals Report 2015* (New York: United Nations, 2015); *Report of the High-level Independent Panel on Peace Operations in uniting our strengths for peace: politics, partnership and people* (New York: United Nations, 2015); *Challenges of sustaining peace: report of the Advisory Group of Experts on the Review of the Peacebuilding Architecture* (New York: United Nations, 2015).

11 *Transforming our world: the 2030 Agenda for Sustainable Development: Resolution adopted by the General Assembly on 25 September 2015* (New York: United Nations, 2015).

12 *One humanity: shared responsibility: Report of the Secretary-General for the World Humanitarian Summit* (New York: United Nations, 2016), 29.

13 Overseas Development Institute, *Time to let go: a three-point proposal to change the humanitarian system*, (2016), 5, available at https://www.odi.org/sites/odi.org.uk/files/resource-documents/10421.pdf.

14 Ibid.

'exogenous "solutions" meeting endogenous "challenges" and "needs"' – [come] into question.[15]

Over recent years, there has been a refocus of state-building policy intervention on the deeper engagement of international agencies and concern with developing new 'bottom-up' approaches to understanding problems and vulnerabilities. Rather than waiting for emergencies to happen, instead there is a deeper, longer-term engagement with ongoing issues, such as extreme poverty amongst the 'most vulnerable'. This is often based on designing indirect forms of intervention for community engagement and empowerment, rather than traditional 'top-down' policy assistance at the level of state institutions.

However, it is not easy to turn 'bottom-up' thinking into a viable form of problem-solving. The essential difficulty appears to be that of the barriers to access and understanding, despite an increasing awareness of the need to differentiate and prioritise by drilling down further (getting more micro-level information) and enabling interventions to be more aligned with complex processes of interaction within and between different local actors and agencies. To overcome this barrier of access, new digital technologies are often held to be key to the reform of international practices,[16] highlighted by the fact that the need to integrate new technological innovations is a constantly recurring theme for international agencies. UN Secretary-General Ban Ki-moon for example, argued that 'Data and joint analysis must become the bedrock of our action. Data and analysis are the starting point for moving from a supply-driven approach to one informed by the greatest risks and the needs of the most vulnerable.'[17]

The difficulty is that it seems that, whatever level of technological drilling down or deeper forms of surveillance and information gathering may be deployed, it is not possible to capture all the potential variables within any given assemblage of interaction. This is particularly problematic

---

15 Ibid., 23.

16 See for example, *A World that Counts: Mobilizing the Data Revolution for Sustainable Development* (New York: United Nations, 2014); Patrick Meier, *Digital Humanitarianism: How Big Data is Changing the Face of Humanitarian Response* (London: CRC Press, 2015); Viktor Mayer-Schonberger & Kenneth Cukier, *Big Data: A Revolution that will Transform how we Live, Work and Think* (London: John Murray, 2013).

17 *One humanity: shared responsibility*, 31; see also Active Learning Network for Accountability and Performance in Humanitarian Action, *Establishing Early Warning Thresholds for Key Surveillance Indicators of Urban Food Security: The Case of Nairobi*, (2016), available at www.alnap. org/help-library/establishing-early-warning-thresholds-for-key-surveillance-indicators-of-urban-food.

for approaches that claim to be inclusive and 'bottom-up'. It appears that any system of data gathering could never be complete or able to grasp processes of interaction in their emergence. It is therefore little wonder that many commentators doubt that the aspirations for digitally enhanced modes of access to relations, in order to fully understand a problem from the 'bottom-up', can be fulfilled.[18] As Nat O'Grady writes, the data categories used for cross-checking risk factors will always be too wide in scope and not targeted enough, thus increasing rather than ameliorating 'the problem of rendering invisible those most vulnerable'.[19]

From Annihilation to Acceptance

By way of preface to the extent of the challenge facing international state-building agencies today, it might be useful to think through the problem speculatively. To this end, *The Southern Reach Trilogy* makes a good starting point. The trilogy of novels by the American author Jeff VanderMeer was first published in 2014: *Annihilation*, *Authority* and *Acceptance*. A film adaptation of *Annihilation* by writer-director Alex Garland was released in 2018. Southern Reach is a military establishment sent to explore and understand the strange 'Area X'. The novels detail some of the experience of sending different missions to the area and how this process of exploration begins to unravel the self-understanding of those involved, without ever gaining any knowledge or control over the area itself. In this mysterious and hypnotic landscape, the military explorers and researchers themselves become 'colonised', losing their illusions and their sanity but never discovering anything that could be called traditional knowledge. The allusions to the inversion of modernist or human-ist understandings of the relationship between the subject and the world – or the bifurcation of culture and nature, which places the human as the knowing, active subject and the world as passive object – are clear in that the two main centres of engagement are the Lighthouse (with its allusions to Enlightenment knowledge) and its inversion, the 'tower'/tunnel, which goes to subterranean depths (which leads to the disorientation and disappearance of the self).

---

18 See Roisin Read, Bertrand Taithe & Roger MacGinty, 'Data hubris? Humanitarian information systems and the mirage of technology', *Third World Quarterly* 37:8 (2016) 1314-1331.
19 Nathaniel O'Grady, 'A Politics of Redeployment: Malleable Technologies and the Localisation of Anticipatory Calculation', in Louise Amoore and Volha Piotukh (eds), *Algorithmic Life: Calculative Devices in the Age of Big Data* (Abingdon: Routledge, 2016), 78.

When the Southern Reach teams cross the border into Area X, their experience of the world becomes 'weird', as they lose their compass on reality and their technological instruments become useless. This experience of the loss of power, control and understanding as leading international institutional experts arrive equipped with their state-building expertise and manuals of 'lessons learned', will be familiar to anyone with experience of state-building missions from Bosnia and Kosovo to Iraq and Afghanistan. The first novel, *Annihilation*, captures the experience of loss of the international state-building narrative of 'progress' and its assumptions that gradually more knowledge can be acquired and problems and barriers overcome:

> Everyone [mission members] had died or been killed, returned changed or returned unchanged, but Area X had continued on as it always had … while our superiors seemed to fear any radical reimagining of this situation so much that they continued to send in knowledge-strapped expeditions as if this was the only option.[20]

The biologist (the leading protagonist) is an expert in 'transitional environments' but, while Area X is a transitioning environment, it is not transitioning in the way imagined by international state-builders. As David Tompkins, in the *Los Angeles Review of Books*,[21] and Joshua Rothman in *The New Yorker*,[22] highlight, Area X can be understood as a 'hyperobject' in the terminology of contemporary speculative realist philosopher Timothy Morton.[23] For Morton:

> Objects are Tardis-like, larger on the inside than they are on the outside. Objects are uncanny. Objects compose an untotalizable nonwhole set that defies holism and reductionism. There is thus no top object that gives all objects value and meaning, and no bottom object to which they can be reduced.[24]

Perhaps it would be useful to think of state-building in these terms: engaging with something that increasingly seems to make more sense if seen as a

---

20 Jeff VanderMeer, *Annihilation* (London: Fourth Estate, 2015), 158–9.
21 David Tompkins, 'Weird Ecology: On The Southern Reach Trilogy', *Los Angeles Review of Books*, 30 September 2014.
22 Joshua Rothman, 'The Weird Thoreau', *The New Yorker*, 14 January 2015.
23 Timothy Morton, *Hyperobjects: Philosophy and Ecology after the End of the World* (Minneapolis MN: University of Minnesota Press, 2013).
24 Ibid., 116.

'hyperobject'. Initially, statebuilders assumed there was a holistic 'top object', which gave value and meaning to what they were doing in building peace, democracy or economic capacities. This top object was the state and its institutions. State-building 'from the top-down' seemed to make a lot of sense, especially in the mid and late 1990s, with international powers highly confident in the attraction of liberal universalism following the collapse of the Soviet challenge.[25] With the perceived failure of state-building from the 'top-down' perspective from the late 1990s onwards, state-building 'from the bottom-up' assumed that, instead of a top object, the focus should be upon a 'bottom object', which could be the (reductionist) key to the state-building project, starting from local and contextual knowledges and approaches and working upwards to achieve international policy ends.

Thus the 'annihilation' in the sub-title of this chapter is not the disastrous experiences of 'top-down' interventions, of 'regime change' and international protectorates, but a much deeper problem of the realisation that international state-builders have no handle on the problems being addressed, for example, that there is nothing to grasp to frame their policy interventions. In addition, there is neither a 'top', to cohere state-building as an external project, nor a 'bottom' from which alternatives can be launched, in terms of basing their legitimacy upon agreement on facts, institutions or actors 'on the ground'. This realisation and its consequences have been much slower to percolate through the levels of international state-building bureaucracy and the 'group think' of the academy, where international institutions have been highly reluctant to think through the implications of the lack of mission legitimacy. The following two sections of this chapter expand on the perception of the problems of 'bottom-up' approaches and the exhausting search for a reductionist ground or foundation. Instead of a 'bottom', international state-builders found themselves in the disorienting experience of infinite complexity, where none of their knowledge or assumptions appeared to provide a useful guide to policy practice. As will be suggested below, very much in line with the speculative 'weird' fiction of the Southern Reach trilogy, the inevitable outcome has been acceptance of the limits of external power and knowledge and, in some cases, a welcoming of the release from modernist or humanist responsibilities for 'problem-solving'.

---

25 Chandler, *Peacebuilding*.

Annihilation

It seems to be logically inevitable that any attempt to start from the perspective of the knowledge and technical mechanisms of international agencies and policy actors will constitute new forms of exclusion and marginalisation. Even if not starting from 'supply-centred' approaches, which assume Western superiority, these state-building approaches nevertheless assume the objective nature of the knowledge of these intervening agencies. In other words, their subject-centred perspectives (of their own role as the active agents, acquiring greater, more varied or more interactive knowledge) is not (as yet) problematised. Thus failures, to know 'the problem' or engage with it successfully, inevitably continue to expose external actors to accusations of being too Eurocentric or Western in their views and not being open enough to the systems and societies in which they are engaged.[26] These forms of criticism cannot be avoided by seeking to develop and innovate technologically, whether it is through Big Data, open-source mapping technologies or other means, as, whatever the nature of the innovation and no matter how extensive its application and how efficient it may be in delivering information, real and complex life can never be adequately captured.[27]

The application of new technologies increasingly reveals the nature of the problem to be different to how it was previously imagined: they reveal communities to be much more differentiated and causal chains are often much more mediated and less linear than previously understood. Acquiring greater knowledge of depth, intricacy and complexity inevitably raises questions about previous knowledge assumptions, and brings attention to the epistemological limitations of external attempts to know societies and processes from the 'bottom-up'.[28] The density is overwhelming. The problem for international actors tasked with policy intervention is that

---

26 Meera Sabaratnam, 'Avatars of Eurocentrism in the critique of the liberal peace', *Security Dialogue* 44:3 (2013), 259–78.

27 Critics have argued that new scanning and mapping technologies may distance humanitarian actors even more from these societies (for example, Tom Scott-Smith, 'Humanitarian neophilia: the 'innovation turn' and its implications', *Third World Quarterly* 37:12 (2016), 2229–51; Mark Duffield, 'The resilience of the ruins: towards a critique of digital humanitarianism', *Resilience: International Policies, Practices and Discourses* 4:3 (2016), 147–65; Rob Kitchin, 'Big Data, New Epistemologies and Paradigm Shifts', *Big Data and Society* 1:1 (2014), 1–12; Claudia Aradau and Tobias Blanke, 'The (Big) Data-Security Assemblage: Knowledge and Critique', *Big Data and Society* 2:2 (2015).

28 See, for example, Peter Finkenbusch, 'Expansive Intervention as Neo-Institutional Learning: Root Causes in the Merida Initiative', *Journal of Intervention and State-building* 10:2 (2016), 162–180.

discussion and reflection upon the epistemological limits of knowledge is bound up with their own external Western positionality.[29]

Contemporary debates over the limits to what international state-builders can achieve thus construct policy interventions as a performative epistemology:[30] the failure of policy-making is seen to directly manifest the limits of a Western way of knowing. This failure is driven by the conflation of epistemological limitations with a Western, Eurocentric or colonial positionality. This positionality is then held to have historically been elitist, hierarchical and exclusionist. The inability to drill down to the required level of depth and grasp the rich interactive density of complex relational processes then gives the lie to state-building claims of objectivity or of epistemic superiority. All interventionist actors (who, by definition, are external agents intervening with instrumentalist intentionality) are caught in the problem of their inability to see the problem in the ways in which it may appear to those more closely involved, despite their claim to be objectively knowing and addressing it. Contemporary political, scientific and philosophical sensitivities necessarily bring international aspirations 'back down to earth', in the knowledge that interveners cannot escape their own socially, politically and technologically mediated frameworks of understanding. It appears that 'bottom-up' approaches cannot step outside of their positionality, even with the nicest and most generous of intentions (or with the most reflexive awareness of the recursive processual nature of assemblages and emergent causality).[31]

Bottom-up or post-liberal state-building interventions,[32] while appreciating non-linear and emergent causality, appear to be unable to overcome the epistemological limits of international policy intervention. State-building as a 'problem-solving' discourse appears trapped in a

---

29 See Pol Bargues-Pedreny, 'From Promoting to De-emphasizing 'Ethnicity': Rethinking the Endless Supervision of Kosovo', *Journal of Intervention and State-building* 10:2 (2016), 222–40.

30 See Andrew Pickering, *The Cybernetic Brain: Sketches of another Future* (Chicago IL: University of Chicago Press, 2010).

31 See Daniela Körppen, Norbert Ropers and Hans J. Giessmann (eds), *The Non-Linearity of Peace Processes: Theory and Practice of Systemic Conflict Transformation* (Opladen: Barbara Budrich Verlag, 2013);

Ben Ramalingam, *Aid on the Edge of Chaos: Rethinking International Cooperation in a Complex World* (Oxford: Oxford University Press, 2013); Cedric De Coning, 'From peacebuilding to sustaining peace: Implications of complexity for resilience and sustainability', *Resilience: International Policies, Practices and Discourses* 4:3 (2016), 166–81.

32 See David Chandler and Oliver P. Richmond, 'Forum: Contesting Postliberalism: Governmentality or Emancipation?', *Journal of International Relations and Development* 18:1 (2015), 1–24.

modernist deadlock, still reproducing 'objective' Western understandings in the attempt to externally 'resolve problems'. However, as is argued here, this experience has opened the possibilities for these limits to be legitimised or worked around. In response to the problems of legitimising knowledge claims, policy innovators are increasingly shifting perspective towards a richer 'posthuman' understanding of knowledge generation.[33] Problems are increasingly recast as ones of phenomenology (of what it is possible to know or have access to) rather than merely epistemology (the question of how knowledge can be gained). Access to and the construction of 'the problem' is thus transformed through the attention to the understanding or perceptions of other agencies or actors.

This shift begins to go beyond the assumptions that the problems are merely ones of epistemology: of extending the knowledge of external actors themselves. Considering how the world might be perceived and questions articulated in different ways, with different tools and techniques, begins to raise questions about the nature of the problem itself.[34] It is at this point that the necessary limits to 'bottom-up' approaches of designing problem-solving interventions appear to become much clearer. Attempting to resolve 'the problem' is then no longer a purely epistemological concern of extending modernist forms of knowledge deeper into social and cultural processes of interaction by fine-tuning techniques of data gathering and breaking down categories of analysis, or speeding up the feedback from digital recording and sensing equipment.

A fundamental gulf opens up between the agency of the international state-builders and the problem itself. Or, rather, the understanding that

---

33 Posthumanist phenomenology is often seen as starting with Thomas Nagel's famous essay, 'What Is It Like to Be a Bat?' (*The Philosophical Review* LXXXIII:4 (1974), 435–50) or with Deleuze and Guattari's popularisation of Jacob von Uexküll's 'ethology' (Gilles Deleuze, *Spinoza: Practical Philosophy* (San Francisco: City Lights, 1988), 124–6), drawing attention to how our perceptions of the world are very different to those of other actors and agencies. This approach pluralises the world, enabling us to see it as constituted through many multiple ways of being, decentring the human as an all-knowing actor. A variety of related approaches – such as speculative realism, object-oriented ontology, actor network theory, new materialism and post-phenomenology – have extended the pluralising perspectives of critical, gender, feminist, black and de-colonial studies to the nonhuman, thus radicalising perspectivism. See, for example, Ian Bogost, *Alien Phenomenology or What it's Like to Be a Thing* (Minneapolis MN: University of Minnesota Press, 2012); Levi R. Bryant, *The Democracy of Objects* (Ann Arbour MI: Open Humanities Press, 2011); Don Ihde, *Postphenomenology and Technoscience: The Peking University Lectures* (Albany NY: State University of New York, 2009); Timothy Morton, *The Ecological Thought* (Cambridge MA: Harvard University Press, 2012); Graham Harman, *Immaterialism: Objects and Social Theory* (Cambridge: Polity Press, 2016).
34 See Karen Barad, *Meeting the Universe Halfway: Quantum Physics and the Entanglement of Matter and Meaning* (London: Duke University Press, 2007).

there is 'a problem' constitutes a fundamental gap between the state-building agency and the society concerned, which is continually apparent when the intervener needs to acquire knowledge to address the problem through providing information and assistance or in terms of knowing more about capacities, choices and needs. Bottom-up interventions emphasised the need for intervention to be 'bottom-up' but, increasingly, it becomes apparent there is no 'bottom' to be found. There is no solid ground for external problem-solving knowledge and expertise, either because this cannot be identified or because this may not exist in any externally appropriable form. With this shift, inevitably, governing and knowing agency necessarily becomes understood as more widely distributed.

The shift to 'bottom-up' or 'problem-centred' approaches seeking to redesign policy interventions appears to have had the additional implication of making societies and 'problems' much more opaque, or rather infinitely more complex, than initially imagined. This forced problems to be increasingly recast as ontological rather than merely epistemological. The point of this distinction is the vantage point or positionality of the knowledge that is required. An epistemological problem can be solved through an expansion of existing frameworks of knowledge, from the subject position of an external actor (in this case, the state-building agency concerned). An ontological shift in perspectives also makes the problem itself less clear; even knowing what 'the problem' is cannot be resolved through such an extension and requires indirect access through the ways of thinking and relating these to the policy target or situation itself. This is demonstrated in the shift in the form of enquiry. The question is then no longer: 'How can we understand more?' or even 'What do they want?' or 'What do they need?', but an entirely different presentation of the question, away from the initial assumption of the existence of 'a' clearly defined problem. Instead, the starting point is more a question of the ways of knowing of others: 'How do they think?' 'How do they see the world?' 'What language do they use?' 'How do they use it?' 'What tools do they use?' 'What instruments or technologies do they use to make sense of the world?' The knowledge sought is how phenomena appear – how information is processed and things are perceived – to others.

In this way, the barriers revealed by the 'bottom-up' approach appear as the barriers of the modern or Western episteme itself, and the renegotiation of intervention as a set of knowledge practices begins to formulate the problem in terms that parallel discussions of posthuman or object-oriented ontologies, where the object of analysis seems to be increasingly obscure and to withdraw or recede from the direct or unmediated view of the external actor.[35] The more that the external state-building agency or actor thinks that it grasps the problem in bottom-up approaches – understands the processes involved, locates the most vulnerable, finds the mechanisms of mediation, interpretation and translation – the more the problem recedes or disaggregates. It is clear that what was mistakenly taken as knowledge of 'the problem' was merely a self-projection of the categories and understandings of the external actor itself. The practical lessons for state-builders, from the experiences of direct forms of intervention in Bosnia, Kosovo, Afghanistan, Iraq, Timor-Leste and many other cases, have been that, rather than going closer to the problem, to addressing causes and removing barriers, the problems appear to move further away, or, more precisely, to have much more relational depth.

Acceptance

The critique of earlier 'top-down' or 'supply-centred' policy approaches – as well as the alternative 'bottom-up' or 'demand-driven' solutions – is precisely that both remain based on projections of Western understandings: of a liberal, modernist or Eurocentric episteme, which makes 'God's eye view' assumptions that the epistemological barriers to problem-solving can be overcome while ignoring the possibility of ontological barriers to knowledge.[36] In the work of object-oriented ontology or speculative realism, this problem of ignoring ontological barriers is often termed 'correlationism', a problematic first coined by Quentin Meillassoux,[37] which is seen to stem from Immanuel Kant's transcendental idealism. Phenomenological barriers to knowledge are not taken seriously, as it is assumed that we never have access to the inner world of experience of other subjects or objects, only to the world as we perceive and experience it, trapped within our own phenomenological world of perception. Thus, problems are always understood epistemologically, within our own set of correlations between the world and ourselves.

Problems thus are always framed as 'problems for us', never constructed, analysed or identified in the ways in which they may appear for other forms of

---

35 See, for example, Harman, *Immaterialism*.
36 See, for example, David Chandler, 'Reconceptualising International Intervention: State-building, "Organic Processes" and the Limits of Causal Knowledge', *Journal of Intervention and State-building* 9:1 (2015), 70–88.
37 Quentin Meillassoux, *After Finitude: An Essay on the Necessity of Contingency* (London: Continuum, 2008), 5–7.

being or ways of existing. The perceived need to overcome or to bypass these limits has been increasingly raised by decolonial approaches[38] and these fit well (in this regard) with the concerns of posthumanist, speculative realist or object-oriented theorists. The key problematic for bottom-up forms of intervention is thus that of not taking alterity or difference seriously enough:[39] the study of different local relations and interactions from the God's eye view of a Western observer or governance agency appears to risk affirming the modernist worldview (of 'phenomenology-of') rather than questioning the hegemonic Western assumptions about the objective or scientific nature of knowledge, for example that the world is single and uniform and only socio-cultural understandings and responses differ.[40]

This reversal of positionality in relation to the problem increasingly links new developments in policy practices with posthuman, speculative or object-oriented approaches. In international policy discourses, bottom-up approaches are fundamentally challenged by the need to go beyond correlationism, beyond merely the projection of a Western external or modernist framing of problems and solutions to think outside these mental constraints. As Meillassoux puts it, this shift can be understood as an exciting challenge of entering 'the great outdoors',[41] no longer forced to be constrained by traditional frameworks of gaining access to problems, but rather to explore other ways of being and knowing.

It cannot be emphasised enough that previous approaches to international state-building are seen to have black-boxed societies, being too little interested in their internal workings and relationships and instead focusing on surface appearances and offering policy advice and assistance on this basis. The opening up of this black box has provided the dynamic that is driving and transforming the design of policy intervention, which increasingly seeks to draw from the rich plurality of the new worlds opened up in the problematisation of a narrow, 'bottom-up' approach. In fact, as

38 See, for example, Walter D Mignolo, *The Darker Side of Western Modernity: Global Futures, Decolonial Options* (London: Duke University Press, 2011); Walter D Mignolo and Arturo Escobar (eds), *Globalization and the Decolonial Option* (Abingdon: Routledge, 2010); Robbie Shilliam, *The Black Pacific: Anti-Colonial Struggles and Oceanic Connections* (London: Bloomsbury Academic, 2015); Sylvia Wynter, 'Unsettling the coloniality of being/power/truth/freedom: towards the human, after man, its overrepresentation – an argument', *CR: The New Centennial Review* 3:3 (2003), 257–337.
39 Matei Candea in Michael Carrithers et al., 'Ontology is Just another Word for Culture: Motion tabled at the 2008 Meeting of the Group for Debates in Anthropological Theory, University of Manchester', *Critique of Anthropology* 30:2 (2010), 175.
40 Holbraad in Carrithers et al., 'Ontology is Just another Word for Culture', 181.
41 Meillassoux, *After Finitude*, 7.

articulated here, it becomes clear that there are two stages of the opening up of the problem. The first stage, external and subject-centred, seeks to drill down, operating within the legacy of the modernist episteme, pluralising the variables and localising the factors (as described above). The second stage begins to shift to a less-modernist framework with a pluralising ontology, speculating upon multiple ways of knowing or perceiving reality or of being in the world.

The attempt to move away from addressing a problem to exploring the ways in which it may appear to others transforms the self-understanding of state-building actors. This shift from a subject-centred position (which assumes a universal or objective perspective) to a posthuman approach is often unclear in the remaking of international discourses of policy intervention because this means dealing with the alien nature, not of objects, but of communities constituted as vulnerable or 'at risk'. Thus Meillassoux's 'great outdoors' becomes recast as an open-ended engagement with the 'other', with the 'local' or with 'grass-roots communities'.[42] The fact that the other can never really be known is not a problem but, on the contrary, positive and enabling, and 'expands possibilities for opening to "new" understandings of difference'[43] where external actors can 'value cultural difference independently of claims to have or know culture, attend directly to the process of constituting culture, and open to other ways of knowing human difference'.[44]

Regardless of the 'bottom-up' terminology through which this shift is recast, the phenomenological framing is the same: policy interventions increasingly start from the perceptions of actors closer to the problem itself and its articulation in its concrete 'local' context. The problem is then posed in terms of the ways of knowing and interacting of the 'local', vulnerable, marginalised or most at risk. Thus the 'project design' shifts from assuming the problem and looking for its solution to a more open-ended enquiry into understanding the perceptions of the 'other' or the ways in which the problem emerges on its own terms.[45] Thus interventions seek to see a more affirmative mode of engagement, exploring how other actors and

---

42 See, for example, Roger MacGinty and Oliver P. Richmond, 'The Local Turn in Peace Building: a critical agenda for peace', *Third World Quarterly* 34:5 (2013), 763–83.

43 Morgan Brigg and Kate Muller, 'Conceptualising Culture in Conflict Resolution', *Journal of Intercultural Studies* 30:2 (2009), 136.

44 Brigg and Muller, 'Conceptualising Culture in Conflict Resolution', 138.

45 See, for example, Aditya Bahadur and Julian Doczi, 'Unlocking resilience through autonomous innovation', *Overseas Development Institute Working Paper* (London: Overseas Development Institute, 2016).

agents see and understand these interactions, grasping the world in its 'ontological multiplicity'. This provides a major challenge in questioning the approach of drilling down to access and open up problems to an external understanding, as these 'bottom-up' approaches are limited by retaining the baggage of the modernist episteme.

Conclusion

This chapter has sought to bring clarity to the discussion of the limits and possibilities of the practices of international state-building intervention, which bring to the surface the difficulty of maintaining the legitimacy of the internationalist imaginary of intervention from a universalist, detached or objective perspective (even if it were possible for state-building interventions to be free from the blinkers of power or ideology). Understanding the 'conditions of impossibility' for traditional or modernist conceptions of international state-building – the inability to legitimate the separations and cuts necessary to demarcate a distinct or separate policy sphere – shines an important light on the frameworks through which policy interventions are understood and contested today. It also suggests that to dismiss posthumanist, speculative realist or object-oriented approaches, as somehow not 'policy relevant', would be to miss the broader context in which both academic and policy processes are evolving.

References

Bahadur, A. and J. Doczi. 'Unlocking resilience through autonomous innovation', *Overseas Development Institute Working Paper*. London: Overseas Development Institute, 2016.

Bargues-Pedreny, P. 'From Promoting to De-emphasizing 'Ethnicity': Rethinking the Endless Supervision of Kosovo', *Journal of Intervention and State-building* 10:2, 2016.

Bennett, C., M. Foley and H.B. Krebs. 'Learning from the past to shape the future: Lessons from the history of humanitarian action in Africa', *Humanitarian Policy Group Working Paper*. London: Overseas Development Institute, 2016.

Björkdahl, A. et al. *Peacebuilding and Friction: Global and Local Encounters in Post-Conflict Societies*. Abingdon: Routledge, 2016.

Brigg, M. *The New Politics of Conflict Resolution: Responding to Difference*. Basingstoke: Palgrave Macmillan, 2008.

Brigg, M. and K. Muller. 'Conceptualising Culture in Conflict Resolution', *Journal of Intercultural Studies* 30:2, 2009.

Chadwick, W., T. Debiel and F. Gadinger. 'Relational Sensibility and the "Turn to the Local": Prospects for the Future of Peacebuilding', *Global Dialogues* 2. Duisburg: Käte Hamburger Kolleg/Centre for Global Cooperation Research, 2013.

Chandler, D. 'Reconceptualising International Intervention: State-building, "Organic Processes" and the Limits of Causal Knowledge', *Journal of Intervention and State-building* 9:1, 2015.

Chandler, D. *Peacebuilding: The Twenty Years' Crisis, 1997–2017*. Basingstoke: Palgrave Macmillan, 2017.

Collinson, S. 'Constructive deconstruction: making sense of the international humanitarian system', *Humanitarian Policy Group Working Paper*. London: Overseas Development Institute, 2016.

De Coning, C. 'From peacebuilding to sustaining peace: Implications of complexity for resilience and sustainability', *Resilience: International Policies, Practices and Discourses* 4:3, 2016.

Debiel, T., T. Held and U. Schneckener. *Peacebuilding in Crisis: Rethinking paradigms and practices of transnational cooperation*. Abingdon: Routledge, 2016.

Duffield, M. 'The resilience of the ruins: towards a critique of digital humanitarianism', *Resilience: International Policies, Practices and Discourses* 4:3, 2016.

Heins, V.M., K. Koddenbrock and C. Unrau. *Humanitarianism and Challenges of Cooperation*. Abingdon: Routledge, 2016.

Körppen, D., N. Ropers, and H.J. Giessmann. *The Non-Linearity of Peace Processes: Theory and Practice of Systemic Conflict Transformation*. Opladen: Barbara Budrich Verlag, 2013.

MacGinty, R. and O.P. Richmond. 'The Local Turn in Peace Building: a critical agenda for peace', *Third World Quarterly* 34:5, 2013.

Millar, G. 'Disaggregating hybridity: Why hybrid institutions do not produce predictable experiences of peace', *Journal of Peace Research* 51:4, 2015.

Nadarajah, S. and D. Rampton. 'The limits of hybridity and the crisis of liberal peace', *Review of International Studies* 41:1, 2015.

Overseas Development Institute, *Time to let go: a three-point proposal to change the humanitarian system*, 2016, available at: https://www.odi.org/sites/odi.org.uk/files/resource-documents/10421.pdf.

Ramalingam, B. *Aid on the Edge of Chaos: Rethinking International Cooperation in a Complex World*. Oxford: Oxford University Press, 2013.

Randazzo, E. *Beyond Liberal Peacebuilding: A Critical Exploration of the Local Turn*. Abingdon: Routledge, 2017.

Sabaratnam, M. 'Avatars of Eurocentrism in the critique of the liberal peace', *Security Dialogue* 44:3, 2013.

Scott-Smith, T. 'Humanitarian neophilia: the 'innovation turn' and its implications', *Third World Quarterly* 37:12, 2016.

2

BUILDING STATES TO BUILD PEACE REVISITED

Empirical Insights from Peace Negotiations in Fragile States

Jan Pospisil[1]

Introduction

The claim that (re)building states is the best way to build peace dominated the peacebuilding debate in the first two decades after the end of the Cold War. This state-building paradigm was a consequence of a focus on working and delivering societal and political institutions which, in turn, resulted from a shift in perceptions of root causes of internal armed conflict from the so-called 'proxy wars' during the Cold War to institutional failure thereafter. However, assessments of how the peacebuilding-state-building nexus plays out in the empirical reality of peace processes are rare. Due to the difficulty of empirical measurement, of defining success and, not least, the complexity of state-building efforts themselves, which consist of a wide variety of practices, the study of state-building is most commonly assessed along a qualitative, case-study based methodology.

1   This research is an output from the Political Settlements Research Programme (PSRP), funded by UK Aid from the Department for International Development (DfID) for the benefit of developing countries. The information and views set out in this chapter are those of the author alone. Nothing herein constitutes the views of DfID.

A majority of these assessments are critical regarding the outcomes of the post-conflict state-building endeavour. One of these studies, conducted in the mid-2000s at the International Peace Institute (IPI) by Charles T. Call and colleagues, lends this chapter its title. The team at IPI critically assessed whether state-building, in the sense of 'building states to build peace',[2] actually results in better post-conflict transitions. The qualitative case study comparison leads to a predominantly pessimistic assessment. The interrelationships between legitimacy, state capacity and security prove to be too complex for state-building interventions to provide for straightforward institution building and subsequent peace. On the contrary, misguided state-building efforts could even exacerbate tensions and result in a relapse into violent conflict.

This chapter revisits the claim made by Call and others regarding the problematic consequences of post-war state-building, based on a newly available data source. The Peace Agreement Access Tool (PA-X) peace agreement database,[3] produced by the Political Settlements Research Programme at the University of Edinburgh, has collected all publicly available peace agreements since the end of the Cold War and coded the agreement texts along 220 categories. Several of these categories refer to provisions dealing with state-building issues. This categorisation, for the first time, enables a broader comparative assessment of state-building endeavours based on the agreements signed to end armed conflict, often by aiming to provide a sustainable institutional setting post-conflict.

Against this background, the chapter aims to address the empirical gap by providing comparative insights into state-building efforts embedded in peace agreements and applying a quantitative approach. However, in contrast to Call et al.'s study, this comparison is not about questioning state-building success. This question remains difficult to answer since stable and widely shared criteria for the success of state-building efforts in post-conflict situations, and reliable data about the implementation of state-building provisions in peace agreements, are scarce. Instead, the chapter examines the trends and preferred methods in post-conflict state-building based on the data peace agreement texts provide. In particular, it investigates the link to power-sharing, which is one of the major approaches used in designing post-conflict transitional frameworks. In

---

2   Charles T. Call, 'Building States to Build Peace? A Critical Analysis', *Journal of Peacebuilding & Development* 4:2 (2008), 60-74; Charles T. Call and Vanessa Wyeth (eds), *Building States to Build Peace* (Boulder CO: Lynne Rienner, 2008).

3   The database is available at http://www.peaceagreements.org.

doing so, the chapter follows the observation of Bell and Pospisil[4] and others[5] that efforts of state-like institutionalisation after conflict often result in a formalisation of political *unsettlement*. The empirical data suggests that state-building efforts in post-conflict environments are, first and foremost, designed to support this formalisation and trigger a long-term and possibly unending transitional process. The promise given by the first generation of state-building research – where the creation of functional statehood would eventually result in normal democratic politics because the institutions would be able to transcend the contestation at the heart of the conflict – has proven to be unfounded.

The remainder of the arguments here are divided thus. Firstly, the chapter provides a critical overview of the trajectories of the state-building discourse following the end of the Cold War. Against this background, a brief discussion of the methodology used in this chapter is given, before engaging more fully with the main empirical insights that state-building provisions inscribed in peace agreements reveal. Finally, some of the wider consequences of these insights for the future practicalities of state-building are analysed.

The Peacebuilding-State-building Nexus and its Challengers

The mutually reinforcing connection between peacebuilding and state-building is a long-standing claim in peace and conflict studies and in International Relations in general. In a historical perspective, the focus on failed states and fragility when analysing the causes and potential solutions of violent conflict represents a specific phase that starts after the end of the Cold War.[6] The intellectual underpinning of peacebuilding at

---

4   Christine Bell and Jan Pospisil, 'Navigating Inclusion in Transitions from Conflict: The Formalised Political Unsettlement', *Journal of International Development* 29:5 (2017), 576–93.

5   World Bank/United Nations, *Pathways for Peace: Inclusive Approaches to Preventing Violent Conflict* (Washington DC: World Bank, 2018), 144–6; Beatrix Austin and Christine Seifert, 'Fourteen regime transitions: what have we learned?', in Hans-Joachim Giessmann and Roger MacGinty (eds), *The Elgar Companion to Post-Conflict Transition* (Cheltenham: Edward Elgar Publishing, 2018), 317–49; Cindy Wittke, *Law in the Twilight: International Courts and Tribunals, the Security Council and the Internationalisation of Peace Agreements between State and Non-State Parties* (Cambridge: Cambridge University Press, 2018).

6   Jan Pospisil and Florian Kühn, 'The Resilient State: New Regulatory Modes in International Approaches to State-building?', *Third World Quarterly* 37:1 (2016), 1–16; see also Christian Büger and Felix Bethke, 'Actor-Networking the Failed State: An Enquiry into the Life of Concepts', *Journal of International Relations and Development* 17:1 (2014), 30–60.

this stage worked as an authorisation for the concept of Western, ideal-type statehood. Several influential writings framed this state-centrism. In peace research, particularly the German tradition pursued such an approach. In the early 1990s, Dieter Senghaas, one of the godfathers of German peace research, developed a so-called 'civilizational hexagon',[7] which rested on the core principles of liberal peace rendered as democratic statehood: the rule of law, democratic representation and the state monopoly of force. In this way, functional and representative statehood allegedly brings about peace and provides the structures to sustain it.

Senghaas' ideas were highly influential in the proliferation of state-building analysis and practice, both within and beyond the German-speaking world. The 'civilising mission' of peacebuilding joined forces with the more security-related assessments prevalent in the Anglo-American world, whereby state failure would not just lead to violent conflict but provided a fertile breeding ground for terrorism and international criminal networks. Consequently, the claim of 'saving failed states'[8] or 'fixing' them[9] started to dominate the concepts, strategies and practices of international intervention. This in turn provoked further reflections in academic thinking, particularly on the role of institutional settings in taming violence,[10] and through popular measurement frameworks such as the 'Fragile States Index' by the Fund for Peace – which began life as the 'Failed States Index' in 2006 – on the root causes of violent conflict and political instability.

From the beginning, the sobering reality of international state-building efforts did not match the conceptual enthusiasm. Effectively, large-scale attempts in the 1990s and early 2000s, starting from Somalia and then progressing through Bosnia-Herzegovina to Afghanistan and Iraq led to, in some cases, highly questionable results. While violence could at least be halted in Bosnia-Herzegovina by the power-sharing framework established by the Dayton Peace Agreement, the other three large-scale efforts continue to see high levels of violence into the present day. These practical challenges have left their mark on the academic debates.

---

7   Dieter Senghaas, 'The Civilisation of Conflict: Constructive Pacifism as a Guiding Notion for Conflict Transformation', in Alex Austin, Martina Fischer and Norbert Ropers (eds), *Transforming Ethnopolitical Conflict: The Berghof Handbook* (Wiesbaden: Springer, 2004), 25–39.
8   Gerald B. Helman and Steven R. Ratner, 'Saving Failed States', *Foreign Policy* 89 (1992), 3–20.
9   Ashraf Ghani and Claire Lockhart, *Fixing Failed States. A Framework for Rebuilding a fractured World* (New York: Oxford University Press, 2009).
10 Francis Fukuyama, *State-Building: Governance and World Order in the Twenty-First Century* (London: Profile Books, 2005); Daron Acemoglu and James A. Robinson, *Why Nations Fail: The Origins of Power, Prosperity, and Poverty* (London: Profile Books, 2012).

Starting with Roland Paris,[11] the productive relationship between state institutionalisation and democratisation was called into question. Institutionalist state-builders argued for an approach later coined as 'institutionalisation before liberalisation' (IBL), whereby, in particular, electoral processes should be postponed until a durable political framework has been set in place and institutionalised. The IBL approach, however, contrasts with international requirements and often those of local stakeholders, which demand a certain level of democratic legitimacy to justify and thereby sustain international support in conflict and post-conflict settings. Despite claims of 'working with the grain',[12] even economically successful cases of state-building based on the IBL approach, such as in Rwanda, are highly contested and challenge international development and peacebuilding policies.

The normative conundrum between institutional functionality and democratic legitimacy was indeed one of the major reasons for the strongly decreasing popularity of state-building in the latter half of the 2000s. The so-called 'Fragile States Principles',[13] an internationally endorsed catalogue of 10 guidelines produced in the framework of the Organisation for Economic Co-operation and Development's (OECD) Development Assistance Committee (DAC), were a striking illustration of this downfall. These principles were aimed at combining the increasing call for contextuality (principle one: 'take context as the starting point') with the dominant focus on state structures in perceived fragile situations (principle three: 'focus on state building as the central objective'). The bottom line was the demand to always work through the systems of the partner country to strengthen weak state capacity.

Besides the normative conundrum, at least two other reasons contributed to the decline of the dominance of the state-building paradigm. First, the epistemic foundations of the state-building enterprise were increasingly called into question. From its beginning, state-building has produced substantial academic critique – it was called, for example, 'empire in denial'.[14] In the latter half of the 2000s, the post-colonial critique meshed with the consequences of the 'local turn' in peacebuilding.[15] The call to

---

11 Roland Paris, *At War's End: Building Peace After Civil Conflict* (Cambridge: Cambridge University Press, 2004).
12 Brian Levy, *Working with the Grain: Integrating Governance and Growth in Development Strategies* (Oxford: Oxford University Press, 2014).
13 These principles can be found at https://www.oecd.org/countries/afghanistan/the10fragilestatesprinciples.htm
14 David Chandler, *Empire in Denial: The Politics of State-Building* (London: Pluto Press, 2006).
15 Roger MacGinty and Oliver P. Richmond, 'The Local Turn in Peace Building: a critical agenda for peace', *Third World Quarterly* 34:5 (2013), 763–83.

embrace context, albeit already inscribed in the Fragile States Principles, became dominant in a way that conceptually did not allow for continuing the positive Eurocentrism that state-building had hitherto represented.

At the same time, the actual subjects of state-building started to become vocal as well. In the framework of the International Dialogue on Peacebuilding and State-building (IDPS) at the OECD DAC, the g7+ group of self-declared fragile states developed and demanded not just a proper seat at the negotiation table with their international state-building partners, but full respect of their state sovereignty. This demand represented a significant shift from debates on 'shared' or 'earned' sovereignty within the state-building paradigm.[16] A considerable achievement by the g7+ group was the development of their own, internationally accepted, measurement framework: the Peacebuilding and Statebuilding Goals (PSGs).

The five PSGs – legitimate politics, security, justice, economic foundations and revenues and services – use a fluid terminology open to interpretation, such as 'inclusive political settlements'. In doing so, they offer a flexible and contextual way to interpret state-building that reaches beyond the ideal type of Western democracies. Their inscribed ambiguity further reflects the practical limits of state-building knowledge.[17] The goals and the five broad stages by which they should be reached – crisis, rebuild and reform, transition, transformation and resilience – also remain vague enough to address a wide range of actors, internally as well as externally.

At the same time, the PSGs represent a continuation of the former conceptual predominance of liberal statehood. Statehood as such remains the main trajectory in post-conflict transformation and the primary orientation of all domestic and international efforts. The ultimate stage of the PSG framework – resilience – is still characterised by functional and democratic statehood, very much defined along the lines of the core principles discussed earlier. These characteristics are transferred into a utopian end-state, whereas all arrangements in between are seen as transitional, which in effect renders 'transition' as an enduring state.

This reflects the second main reason for state-building's decline: the success stories remained mixed, mainly because the core claim that the process of state institutionalisation would be able to transcend the

---

16 Jan Pospisil, 'Unsharing' sovereignty: g7+ and the politics of international statebuilding', *International Affairs* 93:6 (2017), 1417–34.

17 David Chandler, 'Intervention and Statebuilding beyond the Human: From the 'Black Box' to the 'Great Outdoors', *Journal of Intervention and Statebuilding* 12:1 (2018), 80–97.

fundamental contestation at the heart of violent conflicts did not hold. Recent scholarly accounts[18] instead suggest that transitions are not a passing stage in post-conflict settlements, but a rather persistent formalisation of deliberate political unsettlement. The process of state institutionalisation, therefore, hardly ever succeeds in transforming the often-crude power-sharing inscribed in peace agreements into regular politics, but instead perpetuates the transitional process.

Addressing State Fragility While Building Peace: The Evidence

State-building is difficult to measure and challenging to quantify. The written text of peace agreements offers one potential avenue for approaching such measurement, as they contain written, negotiated and, in most instances, agreed stipulations on what should be done in terms of an agenda for post-conflict change. The above-mentioned PA-X, hosted by the Political Settlements Research Programme at the University of Edinburgh, gathers all available peace agreements since 1990: approximately 1,600 documents as part of 136 peace processes, coded along 220 categories, some of them weighted. This data, for the first time, enables comparative quantitative insights into the reality of state-building in the course of peace processes, particularly when using some of the coded categories as proxies for state-building efforts. Peace agreements are negotiated and represent certain political realities. Therefore, they provide a 'living' representation of actual state-building, possibly more than any other form of measurement could. Against this background, the empirical endeavour of this chapter is to explore what state-building looks like in peace agreements and to what extent it is linked to power-sharing.

In doing so, 10 variables coded in PA-X as proxies for the intended strength of state-building in peace agreements are used. These are references to political institutions, a constitution's affirmation, constitutional reform, elections, electoral commissions, political party reform, civil society, public administration, police reform and armed forces reform. All 10 variables are weighted in PA-X, which means that it is possible to distinguish general commitments without any implementation component, from concrete-but-weak commitments, and strong, thoroughly defined commitments.

---

18 Bell and Pospisil, 'Navigating Inclusion in Transitions from Conflict'.

Analysing these 10 variables provides a revealing picture of what and how strong state-building is, both in individual agreements and in peace processes, which usually consist of several separate agreements and cross-process trajectories along a timeline. For the purpose of in-depth analysis, a state-building index is created out of these variables. This index ranges from 0 (none of the 10 proxy variables were present) to an ideal-type score of 3 (all 10 proxy variables were present in maximum strength). An index of 0.5 for a particular year, therefore, means that the average value of a single provision in any individual peace agreement in the given year is at 0.5 (on a scale from 0 to 3). Since a substantial number of peace accords do not refer to state-building at all (for example, ceasefires, implementation agreements), the cross-time comparison of the index shows that any value above 0.3 for a given year is to be considered as high, with top values (for sub-sets of state-building provisions) peaking at a score of 1. The subsequent empirical observations are based on these calculations. They represent state-building as it is reflected in peace agreement texts. While an imperfect proxy, peace agreement texts are still one of the few readily available 'hard' empirical data sources for willingness to engage in and the direction of state-building efforts that can be compared globally.

Approaches to State-building in Peace Agreements

The empirical analysis reveals several insights that are presented in the following section. The first significant empirical insight is that state-building practices broadly follow the trajectories of the international state-building debate as it has been discussed above. In general, state-building measures inscribed in peace agreements get stronger over time. As Figure 2.1 demonstrates, however, state-building interest and activity peaked in the mid-2000s, before flattening out in the latter half of that decade, although at a considerably higher level than before.

The global empirical data chimes with the literature on state-building and state fragility and the international state-building debate, which peaks in these years as well.[19] For example, the OECD International Network on Conflict and Fragility (INCAF) Fragile States Principles were endorsed in

---

19 See Pospisil and Kühn, 'The Resilient State'.

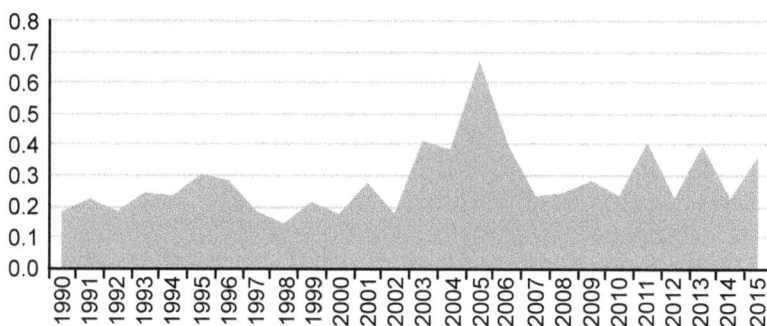

Figure 2.1: Strength of State-building Stipulations in Peace Agreements Across Time (indicator based, per year)
Note: This figure is calculated based on a yearly average on the state-building index per peace agreement in the PA-X database.

2007,[20] following a substantial academic and policy debate on state fragility. The now-famous Fragile States Index by the Fund for Peace was published for the first time in 2006.[21] This shows the close interlinkage between state-building practice and the state-building/fragility debate in policy and academia.

These developments are also driven by substantial peace processes, which also provided the background for further negotiations and academic reflections, such as the local turn. The 1990s were dominated by agreements from the Bosnia and Herzegovina peace process, while a subsequent peak around 2005 is caused by several comprehensive peace processes consisting of a high number of interrelated agreements with strong state-building components occurring at the same time, namely the Comprehensive Peace Agreement (CPA) between Sudan and what later would become South Sudan, the CPA and subsequent constitutional process in Nepal, and peace processes in the southern Philippines/ Mindanao and in Somalia.

A second insight drawn from a more systematic examination of peace agreement texts is what particular types of state-building measures are predominantly negotiated. Figure 2.2 shows the state-building index, calculated per peace process, across the range of coded variables referring to

---

20 Available online at GSDRC http://www.gsdrc.org/document-library/principles-for-good-international-engagement-in-fragile-states-and-situations/.
21 Available online at Fund for Peace https://fragilestatesindex.org.

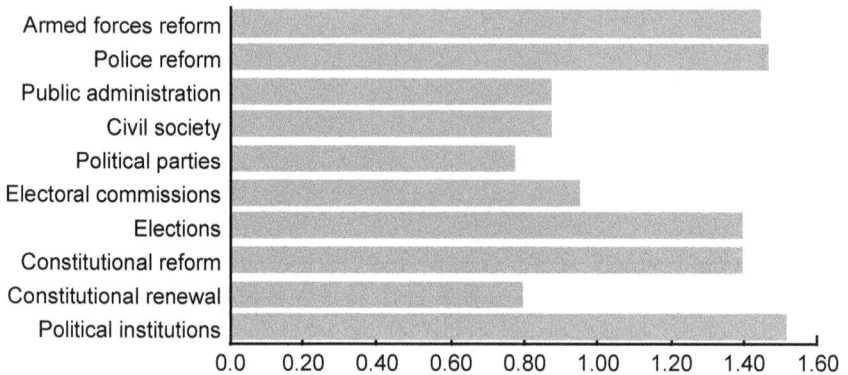

**Figure 2.2: Strength of State-building-related Variables in Peace
Processes**
Note: While this state-building index refers to the same values as in Figure 2.1,
the unit of analysis here is not the individual peace agreement, but the peace
process according to the peace process dyads in the PA-X database. For each of the
processes, the highest value for each variable out of all peace agreements related to
this process is taken. This explains the higher scores.

state-building practices (3 is the highest possible score, 0 the lowest possible
score). This comparison reveals that, across time, the state-building emphasis
is put on two pillars: one pillar is what is commonly referred to as Security
Sector Reform (SSR), namely the strengthening of core security functions and
the reform of the armed forces and the police. Both armed forces reform
and police reform rate above a 1.40 score, which means that they are, on
average, addressed in every peace process with concrete measures that
range beyond a mere rhetorical reference to the necessity of reform.

The second pillar is the work on core institutions of what is perceived
to be essential elements of a functional democratic polity: an implemented
constitutional framework, political institutions and elections. With the only
score above 1.50, political institution-building is still the core pillar of state-
building efforts across time. In contrast, the 'software' of a democratically
constituted policy remains surprisingly weakly addressed. Civil society
scores below 0.90, while political parties register an even lower score below
0.80, making them the weakest component of state reform. A score below 1.0
means that the respective variables are, on average, at best superficially
addressed in peace processes.

Therefore, the comparison shows that the institutionalisation-before-
liberalisation paradigm has taken firm hold in contemporary state-building.

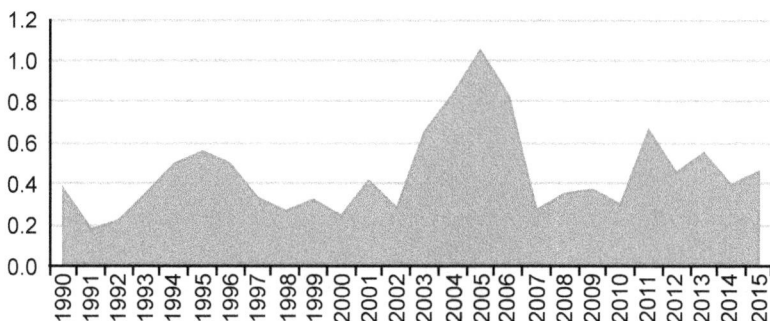

Figure 2.3: Political Institution-building in Peace Processes
Note: The methodology adopted in compiling this figure is as outlined in Figure 2.1.

As will be demonstrated below, the strong interlinkage between institution-building and power-sharing arrangements is responsible for this significant difference. In turn, core elements of what is commonly considered a functional democratic polity – civil society and especially political parties – are only weakly addressed. However, the ability of international state-building support to work on the democratic 'software' needs to be explored further. While practices of institution-building, ranging even to the realm of constitutional reform, can be addressed in a rather technocratic manner, the work on political parties and with civil society is inherently political and, thus, complicated and potentially contested. It remains questionable if the appropriate means exist for international state-building support to progress in these fields.

How strongly state-building is dominated by political institution-building and SSR is further demonstrated by Figures 2.3, 2.4 and 2.5, which show the developments of these key variables over time. The trajectories of institution-building and SSR closely mirror the ups and downs of the state-building enterprise as a whole (as in Figure 2.1). Political party reform and civil society support (Figure 2.4) also mirror the general state-building trend, although at a significantly lower scale. Interestingly, civil society support remains a constant trend throughout post-Cold War history and remains substantially above the level of political party support. Party support, in turn, mirrors the general trend with the expected peaks in the mid-2000s, which is also the only time when it reaches the level of civil society. The influential role of the then-opposition political parties in the Nepali peace

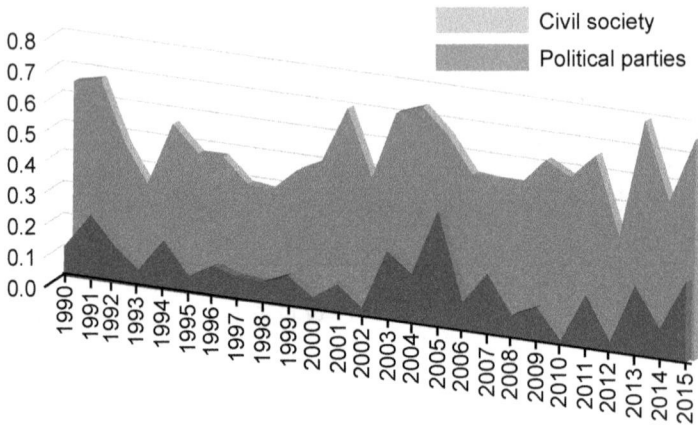

Figure 2.4: Political Party Reform and Civil Society Support in Peace Agreements
Note: The methodology adopted in compiling this figure is as outlined in Figure 2.1.

process explains this trend, since it inflates the numbers around the year of Nepal's CPA, signed in 2006.

The overall composition of state-building in peace agreements does not significantly change over time, which means that the remaining variables reflect the general trend of state-building in peace agreements. This outcome is surprising, given the strong peaking in overall numbers. It also contradicts the trend in state-building conceptualisation within the wider academic literature, especially the call for pronounced contextuality in the aftermath of the local turn and the stronger role in negotiations for fragile states, which was facilitated by the so-called New Deal in the International Dialogue on Peacebuilding and State-building. Instead, state-building seems to have become a 'package deal' in peace negotiations, with an array of common-sense practices that are seen as necessary to be implemented in post-conflict contexts. These patterns amplify the institutionalisation-before-liberalisation agenda.

SSR (see Figure 2.5) in peace agreements is a significant part of the state-building agenda throughout all peace processes since 1990. Interestingly, and somewhat counterintuitively, police reform is a more constant and also stronger element compared with the reform of the armed forces. This is caused by a variety of factors, for example the processes of disarmament, demobilisation and reintegration (DDR) in which members of non-state armed groups become integrated into the police, which

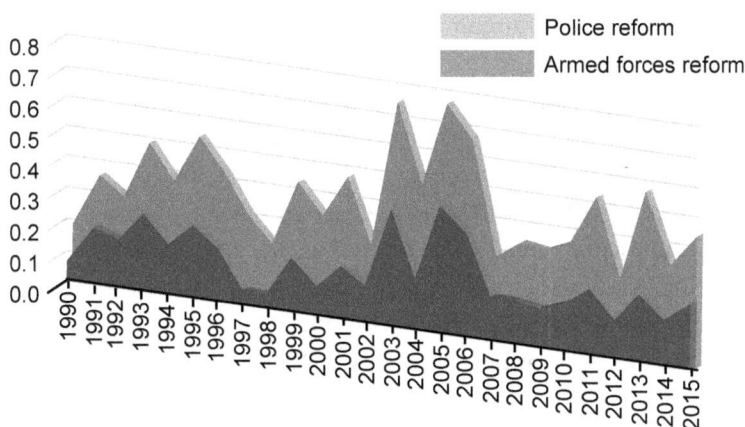

Figure 2.5: Security Sector Reform in Peace Agreements
Note: The methodology adopted in compiling this figure is as outlined in Figure 2.1. Armed forces reform and police reform do not include references to pure combatant reintegration. There needs to be a clearly identifiable reform element attached to any stipulation dealing with armed forces or police.

necessarily results in a substantial reform of the whole police structure, given that non-military police forces tend to be non-existent in such states prior to the peace process.

The Formalisation of Unsettlement: State-building as the Formalisation of Sharing Power

Power-sharing arrangements are one of the central components of peace agreements.[22] They occur in a variety of ways. Besides political power-sharing, economic, territorial and military power-sharing arrangements also feature heavily in peace agreement frameworks. References to power-sharing by consociationalism,[23] a power-sharing concept that guarantees a minimum level of political representation to (for example, ethnopolitical) minorities and social groups, are particularly common.

---

22 Christine Bell, *Political Power-sharing and Inclusion: Peace and Transition Processes*, PSRP PA-X Report: Power-sharing Series (University of Edinburgh: Political Settlements Research Programme, 2018).
23 Timothy D. Sisk, *Power Sharing and International Mediation in Ethnic Conflicts* (Washington DC: United States Institute of Peace, 1996).

Hence, structural arrangements are intended to enable broader democratic representation in a way that representative democracy would not be able to provide.

Often, power-sharing arrangements are meant to be transitional. However, many of these arrangements tend to stick. For example, the Dayton Peace Agreement for Bosnia-Herzegovina, a prototypical power-sharing framework consisting of political, territorial and economic components, was meant to manage the immediate post-war transition by a framework that effectively enshrined the ethnopolitical separation violently enforced during the conflict of 1992–5. Supported by a nominally strong international supervisory regime, it was meant to facilitate the transition to normal politics by introducing electoral democracy and a layer of national politics. Yet, almost a quarter of a century later, the political divisions created by the state institutionalisation based on the power-sharing arrangement are as deep as ever. In all likelihood, the Dayton framework will continue to reproduce the institutional structures for Bosnia-Herzegovina for the foreseeable future.

Empirical comparison shows that this is not a particularity of the Bosnian context: 'While interim transitions often have timetables for reform and elections built into them, in practice these timetables often extend, and the transitional administration often unravels'.[24] State institutionalisation, which is the main component of post-conflict state-building, is according to peace agreement data, strongly linked to the institutionalisation of power-sharing arrangements. In fact, in many instances, these two processes are the same.

Figure 2.6 shows the mean strength of power-sharing arrangements in processes of strong and weak state-building. When looking at the peace process level, the power-sharing components in peace processes with strong state-building are significantly stronger than in those peace processes with weak state-building. This correlation effectively indicates that states in peace processes are much more likely to be strengthened when there is a clear arrangement of how power is to be shared. The motivation behind implementing statehood – to organise and institutionalise the sharing of power – is rather Hobbesian in nature, as it merely aims to underpin a negotiated power-sharing framing with an institutional set-up and, ideally, with the state's monopoly of force. However, such an approach effectively contradicts the 'good governance' rhetoric of development policy actors:

24 Bell, *Political Power-sharing and Inclusion*, 45.

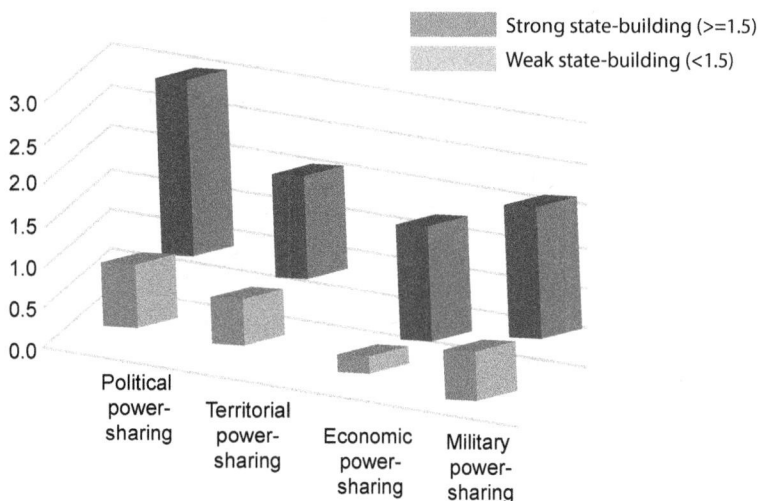

Figure 2.6: State-building and Power-sharing in Peace Processes
Note: For methodological information about how the state-building and the power-sharing indexes are calculated, see the note to Figure 2.2. 'Weak state-building' refers to all peace processes where the state-building index is below 1.5, 'strong state-building' to all peace processes where the state-building index is equal to or above 1.5.

states in peace agreements are not about governance per se, but about *government* in the sense of sharing power – politically, militarily and, to a lesser extent, economically and territorially.

While the strong correlation – although not a cause-effect-relationship – between power-sharing and political institution-building may sound obvious, it still explains why so many peace/state-building processes do not follow the trajectory towards functional, democratic statehood, but instead result in 'formalised political unsettlement'. Political institutions cannot implement democratic governance since they are designed primarily to mitigate the existing political unsettlement. It also explains the neglect of the democratic 'software': political parties and civil society are of course required in post-conflict transitions and have vital roles to play also in the institutionalisation of power-sharing arrangements. However, the post-war transition is not about them. It is about the main conflict actors and their need for accommodation, which the to-be-institutionalised state provides. The institutionalisation of power-sharing by state-building works for all parties because they all win or, at least, do not suffer relational losses vis-à-vis the other parties. Structural transitions towards normal politics, which would require the strengthening of political parties as they would then turn into major actors, are generally referred to in visionary and longer-term language, but not concretely foreseen by the institutional processes set in motion.

Approaching these processes with a lens based on success and failure risks not doing justice to what has – and has not – been achieved. Often, power-sharing arrangements are the only post-conflict deals available at the negotiation table, and they have proven to be able to stop violence and infighting despite the continuation of radical disagreement between the conflict parties. This managerial approach thus needs to be carefully assessed regarding its upsides, even though the resulting situations hardly ever match the expectations that both the wider population and external peacebuilders have.

Conclusion

This empirical comparison of state-building practices as inscribed in peace agreements has revealed three key insights. Firstly, state-building measures since the end of the Cold War are persistently present in peace agreements, with a slightly increasing trajectory and a pronounced peak in the mid-2000s. Secondly, the composition of state-building remains constant, in a way that reflects the institutionalisation-before-liberalisation paradigm. State-building focuses on political institution-building and constitutional settlements, but largely neglects the democratic 'software': in particular political parties but also, to some extent, civil society. Thirdly, this institution-building predominantly serves the formalisation of power-sharing arrangements, which effectively perpetuates mechanisms and arrangements often meant to be transitional.

In general terms, this habitual approach has three advantages: it is easily doable since the main measures are available as practice-based 'toolkits', especially political institution-building (including the related capacity-building) and SSR. International expertise can be provided at comparably low cost and is widely available in all regions of the world. Additionally, these practices and their focus on political structures help to avoid the need for controversial and potentially risky political decision making, which would be required by a more distinct emphasis on political actors. For both diplomatic and development actors, this is a rather convenient option.

When state-building enshrines uncomfortable and unpopular transitional stages in the form of a formalised political unsettlement and is, at the same time, heavily attacked in the wider academic literature, why is it then still such a common and popular approach in peace negotiations that try to design a post-conflict transformational process? The obvious answer to this question is because it serves a multitude of interests, nationally and internationally. It is precisely the enshrining of political unsettlement and the institutionalisation of fundamental political contestation that makes state-building so attractive for conflict actors. In this way, they can sustain their power claims at least to the extent that prevents the other side winning along its default positions.

Second, statehood – or, at least, its image – is a desirable goal not only for the conflict actors, but also for external mediators and supporters and the international community and the international system as a whole. Statehood is a prize because of the international legitimacy it guarantees (and, for example, potential external funding down the line) which, in turn, also stabilises the current international system in the given region. Not having states is a structural danger that reaches far beyond the actual spatial dimension of a non-state. It contests the structure as a whole. Against this background, the failure of the so-produced (or reproduced) statehood to deliver the required political and social goods to its citizens, as any transformational process would demand, is an unfortunate side-effect.

Finally, the relationship between knowledge production and policy practice in state-building is in dire need of more thorough reflection. The empirical comparison in this chapter has shown that these two layers are closely interwoven, although both retain their particularities. The timeline suggests, however, that this does not mean that academic evidence is implemented by policy. It is the other way around: policy problems and policy practice are reflected and theorised by academia in the aftermath of their attempted implementation. Against this background of circular referencing, academic research and policy making need to be careful not to accommodate themselves in a tautological relationship.

References

Acemoglu, D. and J.A. Robinson. *Why Nations Fail: The Origins of Power, Prosperity, and Poverty.* London: Profile Books, 2012.

Austin, B. and C. Seifert. 'Fourteen regime transitions: what have we learned?', in H-J. Giessmann and R. MacGinty (eds). *The Elgar Companion to Post-Conflict Transition*. Cheltenham: Edward Elgar Publishing, 2018.

Bell, C. *Political Power-sharing and Inclusion: Peace and Transition Processes*, PSRP PA-X Report: Power-sharing Series. University of Edinburgh: Political Settlements Research Programme, 2018.

Bell, C. and J. Pospisil. 'Navigating Inclusion in Transitions from Conflict: The Formalised Political Unsettlement', *Journal of International Development* 29:5, 2017.

Büger, C. and F. Bethke. 'Actor-Networking the Failed State: An Enquiry into the Life of Concepts', *Journal of International Relations and Development* 17:1, 2014.

Call, C.T. 'Building States to Build Peace? A Critical Analysis', *Journal of Peacebuilding & Development* 4:2, 2008.

Call, C.T. and V. Wyeth (eds). *Building States to Build Peace*. Boulder CO: Lynne Rienner, 2008.

Chandler, D. *Empire in Denial: The Politics of State-Building*. London: Pluto Press 2006.

Chandler, D. 'Intervention and Statebuilding beyond the Human: From the '"Black Box"' to the '"Great Outdoors"', *Journal of Intervention and Statebuilding* 12:1, 2018.

Fukuyama, F., *State-Building: Governance and World Order in the Twenty-First Century*. London: Profile Books, 2005.

Ghani, A. and C. Lockhart. *Fixing Failed States. A Framework for Rebuilding a fractured World*. New York: Oxford University Press, 2009.

Helman, G.B. and S.R. Ratner. 'Saving Failed States', *Foreign Policy* 89, 1992.

Levy, B. *Working with the Grain: Integrating Governance and Growth in Development Strategies*. Oxford: Oxford University Press, 2014.

MacGinty, R. and O.P. Richmond. 'The Local Turn in Peace Building: a critical agenda for peace', *Third World Quarterly* 34:5, 2013.

Paris, R. *At War's End: Building Peace After Civil Conflict*. Cambridge: Cambridge University Press, 2004.

Pospisil, J. '"Unsharing" sovereignty: g7+ and the politics of international state-building', *International Affairs* 93:6, 2017.

Pospisil, J. and F. Kühn. 'The Resilient State: New Regulatory Modes in International Approaches to State-building?', *Third World Quarterly* 37:1, 2016.

Senghaas, D. 'The Civilisation of Conflict: Constructive Pacifism as a Guiding Notion for Conflict Transformation', in A. Austin, M. Fischer and N. Ropers (eds), *Transforming Ethnopolitical Conflict: The Berghof Handbook*. Wiesbaden: Springer, 2004.

Sisk, T.D. *Power Sharing and International Mediation in Ethnic Conflicts*. Washington DC: United States Institute of Peace, 1996.

Wittke, C. *Law in the Twilight: International Courts and Tribunals, the Security Council and the Internationalisation of Peace Agreements between State and Non-State Parties*. Cambridge: Cambridge University Press, 2018.

World Bank/United Nations. *Pathways for Peace: Inclusive Approaches to Preventing Violent Conflict*. Washington DC: World Bank, 2018.

3

THE SELECTIVE NATURE OF INTERNATIONAL RESPONSES TO FRAGILE, FAILING AND FAILED STATES

Jacob Thomas-Llewellyn

Introduction

While the lexicon of 'fragile', 'failing' and 'failed' states creates a broader spectrum of some semblance of defined crisis, these vulnerable states all present at least one common theme: they are phenomena that potentially present a destabilising effect on the contemporary international system. Successful states that participate in and conform to this system are themselves diverse, in history, power status, character (culture) and capital (economic, diplomatic and military). It could therefore be considered obvious that approaches to dealing with emerging risks and threats from fragile, failing and failed states will be divergent and contested.

Poorly executed interventions in vulnerable states, from those of the early 1990s through to 2017, suggest the US and its allies in Europe have failed to resist the rigours of 'intervention fatigue'.[1] In the majority of cases, successful intervention has proven extremely challenging to deliver and/or maintain, even as the intervening powers in question seem

---

1   Lothar Brock et al., *Fragile States: Violence and the Failure of Intervention* (Cambridge: Polity Press, 2012), 104.

desperate to discover a 'quick fix' to reduce the time taken on the task, stem the financial outlay and protect their reputation. The result has been the broad adoption of what might be called a tailored minimalist approach to interventions, which has generally manifested itself as a deliberately selective strategy where actions primarily serve the national interests of the intervening powers. According to Bellamy and Wheeler, 'states almost always have mixed motives for intervening and are rarely prepared to sacrifice their own soldiers overseas unless they have self-interested reasons for doing so ... this means that genuine humanitarian intervention is imprudent because it does not serve the national interest ... it [also] suggests that the powerful only intervene when it suits them to do so'.[2] From Somalia in 1992 to recent operations in volatile areas of West Africa, it is evident that this 'realist' approach has largely manifested itself in the adopted intervention policies.

For effective and timely action to address state fragility and failure, there must exist robust contingency plans coupled to an enduring political will on the part of the international community. Contemporary international approaches have proven to be highly selective, with several recurring contributory factors undermining a credible response. One major factor is the uncompromising and unrealistic timeframe adopted to deliver solutions for states and regions that are spiralling into fragility, failure or even the complete disintegration of self-governance. National leaders have consistently delivered sweeping public statements that provide little insight into how necessary rebuilding is to be achieved. Following the initial coalition bombing of Libya in spring 2011 by the UK, US and France, prior to NATO taking responsibility as part of Operation Unified Protector (OUP), President Obama declared that the US had been supporting the people of Libya:

> As they build a future that is free and democratic and prosperous ...
> Today the Libyans are writing a new chapter in the life of their nation.
> After four decades of darkness, they can walk the streets, free from a
> tyrant ... And here at the UN, the new flag of a free Libya now flies
> among the community of nations.[3]

2   Alex J. Bellamy and Nicholas J. Wheeler, 'Humanitarian Intervention in World Politics' in John Baylis, Steve Smith and Patricia Owens (eds), *The Globalization of World Politics: An Introduction to International Relations* (Oxford: Oxford University Press, 2011), 514.
3   White House Press Office 2011, 'Remarks by President Obama at High Level Meeting on Libya', available at https://obamawhitehouse.archives.gov/the-press-office/2011/09/20/remarks-president-obama-high-level-meeting-libya

Yet, within a year of this statement, Libya's second principal city, Benghazi, was declared one of the most dangerous areas in the world and, by September 2012, the city had been reduced to a war zone by rival militias. To emphasise the fragility of the situation, the US Ambassador, Christopher Stevens, was killed during a surprise attack that month. Libya remains one of the most 'fragile' states in the world, according to the Fragile States Index (FSI), moving from 111th to the 23rd most fragile between 2010 and 2017.[4] Libya is not an isolated case and there is a continuing failure on the part of the international community, in particular the Western powers, to look beyond the initial intervention phase and appreciate that, for stability to be enduring, long-term, multifaceted commitments are required.

By focusing on a range of historical and contemporary case studies, it is possible to identify why international approaches to the challenge of state fragility and failure persist in being selective in nature. This chapter will adopt a chronological approach, taking the 'humanitarian' interventions of the 1990s as a starting point. Since the debacle of Rwanda (1994), there have been several opportunities for the international community to take decisive action in response to state fragility and failure, including during the Ebola outbreak in West Africa (2014) and the persisting violence in Myanmar. Yet, in all of these cases, a combination of factors contributed to the employment of minimalist approaches which, to a large extent, remain poorly coordinated, lack underlying political will and thus deliver only sub-optimal end states. The case studies discussed here demonstrate that international responses remain inadequate as a result of information and resource overstretch, the constraints of international law, a failure to determine long-term commitments and the all-pervading influence of self-interest.

Case Study 1: Rwanda 1994

In 1988, the Russian political propagandist Georgi Arbatov declared to the West that, with the decline of the US's principal Cold War rival, 'we are going to do a terrible thing to you … we are going to deprive you of an enemy'.[5] Arbatov accurately captured the dilemma that the US and its European allies would face following the demise of a bipolar world a few

---

4   Fund For Peace 2018, 'Libya', available at http://fundforpeace.org/fsi/country-data/. See also the discussions by Islam Goher in Chapter 7 of this volume.
5   Quoted in E. Kominsky, 'Amongst the Chatter, America Burns', *Harvard Political Review*, 2011, available at http://harvardpolitics.com/books-arts/amongst-the-chatter-america-burns/.

years later. Without a powerful state rival to focus on, the Western states were resigned to identifying new security challenges and, since 1989, they have embarked on ever-more ambitious missions to 'stabilise' those areas of the globe deemed to pose risks or threats to the international system. Yet, in the process of identifying these threats in the post-Soviet world, Gen Timothy Cross has rightly noted that the liberal Western bloc has 'lost clarity and … does not know what its strategic intent is'.[6]

The early 1990s, which were optimistically described as a potential 'golden age'[7] for UN interventions, were actually marked by repeated failures on the part of the international community and Western states in particular. The major setbacks that the US experienced in Somalia in the wake of the crisis in Mogadishu in 1993 set the tone for many future interventions, with television images of American bodies being dragged through the streets not only undermining American political will, but also impairing the future resolve of other Western states.[8]

When violence broke out on the streets of the Rwandan capital, Kigali, in April 1994, it was widely assumed that another 'tribal' war had erupted. Such assumptions are unsurprising, as few people across the globe could effectively appreciate its political and cultural complexities. During an interview with a British Army commander deployed to Rwanda in July 1994, it transpired that, while boarding their military transport, planning staff were 'consulting Encyclopaedia Britannica to find out where Rwanda was'.[9] The international community and representatives of the UN had appeared to be completely unsighted by the scale and organisation of the violence, yet the ethnic/tribal tensions as they were portrayed in the Western media had been fermenting for almost a century – at least since Imperial Germany first colonised the region in 1898.[10]

From the early twentieth century, colonial powers – including Belgium and France – had continued to exercise influence over Rwanda. This external manipulation contributed to the tensions between the Tutsi and Hutu communities. Unwittingly, the issuing of identity cards by the Belgians in the 1930s not only codified racial differences, but created

6   Author's personal communication with Gen Timothy Cross, April 2018.
7   See Richard Falk, *Achieving Human Rights* (London: Routledge, 2009), 166.
8   On this see Jean Baptiste Jeangene Vilmer, 'Does the 'CNN Effect' exist? Military Intervention and the Media', IGN Global, 28 June 2012 [in French], available at https://www.inaglobal.fr/en/ideas/article/does-cnn-effect-exist-military-intervention-and-media.
9   Author's interview with former British army officer, 2018.
10 See Ben Kiernan, *Blood and Soil: A World History of Genocide and Extermination from Sparta to Darfur* (London: Yale University Press, 2007), 555.

a mechanism by which each community could easily identify its tribal neighbours.[11] The dominant Tutsi minority, with support from the resident Belgians, utilised the identity cards to exclude Hutus from any form of government decision-making bodies. As a result of a 1959 revolt, the state witnessed a complete reversal in the relationship between the Belgian authorities and the Tutsi community. To maintain some semblance of control, the Belgians switched their support to favour the Hutu majority. The security of the Tutsis was then further undermined in 1961, when Rwanda formally received its independence and international oversight was lost. In 1973, the Hutu leader, Juvenal Habyarimana, ascended to the Rwandan presidency following a *coup d'état* and instigated the adoption of a hard-line racial doctrine through an extremist inner leadership that eventually in turn instigated the genocide two decades later. To underpin their control, the Hutu government initiated a rapid militarisation programme that required the courting of friendly states to secure favourable lines of credit for the procurement of large arms inventories.[12] Although it is challenging to identify a definitive catalyst for the eventual genocide, it was these deliberate, overt policies that set the conditions for the Hutu government to initiate a violent and destructive campaign against the Tutsi minority.

Those who witnessed the events leading up to the genocide might argue that, with hindsight, it is easy to criticise the international community's apathetic response. However, as a 1995 report produced by the African Rights advocacy group noted, there were 'numerous warning signs for the apocalypse that was unleashed. This is not a matter of hindsight'.[13] The international reaction to the deteriorating situation prior to 6 April 1994 (when the killings commenced) fell well short of a credible and effective response. Initially, UN members were both ignorant and misguided in their thinking and, as a consequence, were spectacularly slow in formulating a strategy that offered protection to the Tutsi population. When members eventually took heed of the reports in their possession they seemed to enter into a state of confusion. They immediately *reduced* the size of the existing United Nations Assistance Mission for Rwanda (UNAMIR) in Kigali and overcompensated for their ineffectiveness by deploying personnel, equipment and humanitarian aid into a volatile situation in an uncoordinated manner with no agreed end state.

---

11 Ibid., 555.
12 African Rights, *Rwanda: Death, Despair and Defiance* (London: African Rights, 1995), 66–8.
13 Ibid., 1103.

Before all this, significant international diplomatic and economic pressures had brought Habyarimana to the negotiating table in July-August 1993. To a large extent, all parties involved (apart from the Hutu extremists) were determined to establish a settlement to the ongoing civil conflict, yet the terms of what would become the Arusha Accords were entirely unacceptable to the hard-line elements of Habyarimana's party.[14] In addition to creating new democratic elections, the accords called for a restructuring of Rwanda's armed forces, which would enable the Tutsi community to make up a larger percentage of the national army and allow Tutsi officers to hold key positions. When considered with the fact that Rwanda's regional neighbours, such as Uganda, were keen for exiled Tutsi communities to vacate their temporary homes there, a situation manifested itself that had dire human consequences. The drafters of the Arusha Accords completely misread the situation, either out of sheer ignorance or as a result of being misled.

Although the scale of the violence in April 1994 took the international community by surprise, there were several indicators that suggested a violent confrontation was highly likely to occur, and it can only be assumed the UN and the member states on its Security Council (UNSC) either lacked sufficient capacity or will to process such information in a timely and effective manner. Across the UN diplomatic corps, state representatives and the secretariat, there were several attempts to rationalise their inactivity, primarily blaming the paucity of accurate information. In large part, the bureaucratic culture of the UN stifled representatives from initiating a decisive and timely response. As Michael Barnett has observed, 'something about the culture of the UN could make non-intervention not merely pragmatic but also legitimate and proper – even in the face of crimes against humanity'.[15] In the face of the growing crisis on the ground, the then UNAMIR force commander, General Romeo Dallaire, pleaded with UN headquarters in New York to heed his reports that local Hutu militias were systematically being mobilised and stockpiles of arms were being distributed for deliberate and large-scale acts of violence. In response, UN headquarters delivered 'only rhetoric in the hope that rhetoric represents its own consolation'.[16] As a result, and in the wake of the Rwandan

---

14 Nicholas Wheeler, *Saving Strangers: Humanitarian Intervention in International Society* (Oxford: Oxford University Press, 2000), 212.

15 Michael Barnett, *Eyewitness to Genocide: The United Nations and Rwanda* (Ithaca NY: Cornell University Press, 2002), xi.

16 Ibid., xii-xiii.

president's death in an unexplained plane crash, around five people would be systematically, deliberately and continuously killed every minute during the genocide which lasted from 6 April to 19 July 1994.[17] To confront this mass slaughter, General Dallaire was left with a wholly overmatched force of 270 personnel, with limited logistical support and transport and thus no credible capability to confront and restrain either Hutu government forces or the rural militias when order began to break down.[18]

To understand the weak status of UNAMIR, it is important to appreciate the composition of the force and how far the international contributions fell short of Dallaire's initial requests. Arguably, there was no credible show of force, as the initial deployment of 2,548 personnel was eventually reduced to 450.[19] Dallaire had expected four utility helicopters, eight military helicopters and 22 Armoured Personnel Carriers (APCs) to form the core of a quick reaction force. In reality, UNAMIR possessed no helicopters and only five elderly Czech BTR-80 APCs. When the APCs arrived in March 1994, it was discovered that they were 'without tools, spare parts, mechanics or manuals and [with] limited ammunition'.[20] The state of Dallaire's force not only highlighted the fact that no members of the UN were willing to provide credible equipment, it also exposed the chaotic nature of the UN's own logistical planning at the time. It remains unclear if this weakness has been properly and effectively rectified by the UN even now, but the states that retain credible global military response capabilities seem challenged to maintain their capacity for comprehensive interoperability and to operate willingly and effectively within a UN framework. This calls into question the ongoing level of political willingness, beyond rhetorical exhortation, to effectively respond when order within states breaks down, even in such a dramatic fashion and with such brutal consequences, as was the case in Rwanda.

As well as the poor state of UNAMIR equipment, the mission also lacked adequate logistical support. International contingents, including Ghanaian and Bangladeshi detachments, arrived with only the equipment they were carrying and without food, fuel or stocks of ammunition.[21]

---

17 Ibid., 3.
18 General Romeo Dallaire, *Shake Hands with the Devil: The Failure of Humanity in Rwanda* (Toronto: Vintage Canada, 2004), 177.
19 Dallaire, *Shake Hands with the Devil*, 117.
20 Linda R. Melvern, *A People Betrayed: The Role of the West in Rwanda's Genocide* (London: Zed Books, 2006), 85.
21 Dallaire, *Shake Hands with the Devil*, 123–4.

UNAMIR did not possess the contingency stocks or a robust supply chain to sustain itself. As a consequence, ammunition stocks were scarce throughout the mission, with UN troops compelled to 'scrounge'.[22] In this perceived non-strategic backwater, the UN had struggled to develop a compelling geopolitical narrative that could secure credible military capabilities and there is scant evidence the situation has changed in the contemporary political environment. The irony is that, while there was a paucity of credible capability allocated to General Dallaire, there was no shortage of misguided minor bureaucratic regimes that consumed the time of those deployed on the ground while doing little to alleviate the suffering. At times, it seemed a case of activity over accomplishment. Dallaire's personal recollections expose the attritional administrative battle he was consigned to fight, exacerbating his already serious operational challenges during a rapidly deteriorating security situation. According to Linda Melvern, most of Dallaire's time and energy 'went into trying to sort out a logistical nightmare'.[23] Unsurprisingly, when the genocide began, UNAMIR was powerless to effectively intervene, and it was only after the killing had subsided with the intervention of the Rwandan Patriotic Front (RPF) that the international community focused on delivering a more credible response. This torpidity in the UN system can partly be explained by the spectre of Somalia the year before, which continued (and to an extent still continues) to haunt the UN leadership and Western members of the UNSC.[24]

In contrast to the woeful state of preparedness for protecting Rwandan citizens from the failings of their state, the level of international self-protection was notably higher. Following the fatal ambushing of ten Belgian peacekeepers on 7 April 1994, the public backlash in Belgium forced UN representatives to re-evaluate their position in Rwanda. Only five days after the deaths of its peacekeepers, Belgium announced that it would be withdrawing its military contribution and, according to the findings of African Rights, only two minutes after the last of the Belgian contingent withdrew from their compound, the Rwandans who had sought shelter with these troops were systematically killed by the *Interahamwe* (local Hutu militia) and elements of the presidential guard.[25] Within a wider context, inside UN headquarters, the consensus was on minimising losses to

---

22 Ibid., 177.
23 Melvern, *A People Betrayed*, 86.
24 Walter Clarke and Jeffrey Herbst, 'Somalia and the Future of Humanitarian Intervention', *Foreign Affairs* 75:2 (1996), 70.
25 African Rights, 1113.

international personnel. Britain's then-ambassador to the UN, Sir David Hannay, called upon it to consider the events in Mogadishu the previous year and the potential risk to both personnel and the reputation of the UN, in stressing the importance of scaling back the UN's already limited commitment in Rwanda.[26]

Ultimately, the international community's will to act in Rwanda was found severely wanting, and it was only *after* the genocide had subsided and the UN's reputation was on the line that international aid and effective troop deployments began to materialise. Ultimately, a number of factors coalesced to produce a deficient international response that effectively left an under-resourced UN ground force stranded in the midst of an extremely volatile situation the international community struggled to even understand. The genocide in Rwanda proved to be a painful and depressing legacy for the continent and the international community. Even so, states with the capability to intervene continue to distance themselves from fragile situations and, as demonstrated below, an international response is only initiated when a deteriorating situation threatens to impinge on the security interests of leading Western states.

Case Study 2: The West Africa Ebola Crisis 2014

According to Lucy Westcott, 'natural disasters like floods and earthquakes usually prompt a generous outpouring of resources and direct intervention from aid organisations and concerned states, but fear of the unknown and lack of expertise in Ebola paralysed most aid agencies and donors'.[27] Similarly to events in Rwanda 20 years before, the Ebola virus disease (EVD) that struck several states in West Africa in early 2014 caught the wider international community in a state of unpreparedness, suggesting on the face of it an insufficient focus on ensuring a more consistent and constructive focus on development in Africa, either by leading Western states or key international organisations and agencies.

The initial international response proved ineffective and, indeed, arguably contributed to the spread of the virus. According to the director of Médecins Sans Frontieres (MSF), Christopher Stokes, the 2014 Ebola

---

26 Wheeler, *Saving Strangers*, 221.
27 Lucy Westcott, 'Doctors Without Borders Slams Slow International Response to Ebola', Newsweek 2015, available at http://europe.newsweek.com/doctors-without-borders-slams-slow-international-response-ebola-316089?rm=eu.

outbreak was, on first sight, 'a perfect storm; [a] cross-border epidemic in countries with weak public health systems that had never seen Ebola before'. He added, however, that 'this is too convenient an explanation. For the Ebola outbreak to spiral this far out of control required many institutions to fail and they did, with tragic and avoidable consequences'.[28]

Initially, the World Health Organization (WHO) failed to effectively comprehend the reports from MSF and other humanitarian agencies, which indicated that an epidemic was unfolding. Through a combination of fear, unclear information on the ground and the challenge of mobilising manpower and finance, the approach of the international community, local ministries and Non-Governmental Organisations (NGOs) proved sluggish and poorly coordinated. The crisis also demonstrated several negative aspects of the modus operandi of NGOs and the sometimes highly selective nature of their leaderships in the pursuit of both funding and reputation.

Unlike previous outbreaks, the disaster of 2014–16 witnessed the rapid spread of the disease not only among local populations, but cross-border from Guinea to Sierra Leone and into Liberia.[29] Due to the weakness of their healthcare systems and the paucity of border controls – state fragility, if not failure, at multiple levels of governance – the states affected were extremely vulnerable. The number of victims was also unprecedented, as shown in Table 3.1.

Matters were complicated not only by the speed of the spread of infection across the West African region, but also the complex debilitating effect of the fear that the virus generated, combined with the debilitating effects of corruption in host states. In the initial stages of the crisis, the WHO also lacked effective leadership and was slow to declare an emergency for fear of the potential fallout for the West African economy (which is, to be fair, itself a contributor to state fragility in the region).[30] The international community then further compounded the crisis by initiating sweeping measures to deny any flights to the region which, while seeking to limit the spread of the disease to the wider international arena, had the counter-productive impact of ultimately hampering aid deliveries to those already infected or at risk.[31] To protect themselves and prevent a wider

---

28 Westcott, 'Doctors Without Borders Slams Slow International Response to Ebola'.
29 World Health Organization 2018, 'Ebola Virus Disease', available at http://www.who.int/en/news-room/fact-sheets/detail/ebola-virus-disease.
30 Westcott, 'Doctors Without Borders Slams Slow International Response to Ebola'.
31 Laurie Garrett, 'Ebola's Lessons: How the WHO Mishandled the Crisis', *Foreign Affairs* 94:5 (2015), 104.

Year	Country	Cases	Deaths
1976	DRC	318	280
1976	Sudan	284	151
1995	DRC	315	254
1996	Gabon	91	66
2001–2003	Congo	237	201
2011–2012	Uganda	32	22
2014–2016	Guinea	3811	2543
2014–2016	Liberia	10675	4809
2014–2016	Sierra Leone	14124	3956

Table 3.1: Ebola Outbreaks 1976–2016
Source: World Health Organization, 2018.

crisis, therefore, the end result of this aspect of the international response was thus to exacerbate regional failings.

The failure of the initial response was further undermined by the high levels of corruption that significantly impinged on the relief effort, as can be seen when considering the UK's response to the outbreak in Sierra Leone in August 2014, where it confronted a 'lack of a coherent response' that had 'resulted in pervasive, uncontrolled infectivity'.[32] Sierra Leone at this time was registering 200 new cases of infection per day between September–November and, if WHO predictions were to be believed, would reach 20,000 cases per week by December without a properly coordinated programme that isolated victims and provided adequate decontamination facilities.[33] As part of Operation Gritrock, the UK initiated a tactical plan named 'Western Area Surge' in an effort to quarantine Sierra Leone's capital, Freetown. According to personnel who took part in the operation, corruption undermined their efforts at every turn. In many cases, the process of screening at checkpoints suffered as a result of embezzlement, which resulted in medical personnel not receiving their salaries.[34] The prominence of corruption, especially in chaotic situations, is not a new

---

32 Steve McMahon, 'Op GRITROCK – Fighting Ebola and the Legacy of Defence Engagement', *British Army Review* 164 (2015), 9.
33 McMahon, 'Op GRITROCK', 10.
34 Sinead Walsh and Oliver Johnson, *Getting to Zero: A Doctor and a Diplomat on the Ebola Frontline* (London: Zed Books), 358.

phenomenon among fragile developing states, as 'the injection of large amounts of resources into resource-poor economies where institutions have been damaged or destroyed increases the opportunities for the abuse of power'.[35]

In addition, however, embezzlement by officials and staff was a factor across several aid agencies, including the International Red Cross (IRC), which reported the loss of $5million as a result of personnel selling supplies on the black market and the creation of salaries for non-existent workers.[36] These examples of corruption within the NGO community – in some ways replicating the problems of government agencies they are trying to assist – complicated an already broad cultural problem that pervades both responders and recipients.[37] The events in Sierra Leone and its neighbours demonstrate that, for future operations, credible international assurance measures to combat corruption must be incorporated within aid efforts; it may even be necessary to devise mechanisms for limiting the disbursement of aid and charity in such a way that it does not overwhelm already fragile capabilities within a recipient state.

In spite of the fact that numerous NGOs provide crucial support for aid relief operations, the wider culture of NGOs is something that is increasingly identified as a potential hindrance to effective aid distribution. Reflecting on the humanitarian crisis in Goma in 1994, for example, Duncan noted dryly that 'NGOs seemed to have an inexhaustible supply of T-shirts with their coloured logo on display, some were more keen on jockeying for TV-friendly locations than on tackling humanitarian problems'.[38] The commercial incentive to be seen to act – which may reflect wider incentives within the international community as it selects where, how and when to intervene – has led to increased tension between NGOs and other actors involved in post-crisis stabilisation, notably the military. Recalling the events of the 1999 Kosovo refugee crisis, General Tim Cross noted that

35 Transparency International 2014, 'Saving Lives: Fighting Corruption in Humanitarian Assistance', available at https://www.transparency.org/news/feature/saving_lives_fighting_corruption_in_humanitarian assistance
36 BBC News 2017, 'Red Cross Apologises for Losing $5milion of Ebola Funds to Fraud', available at http://www.bbc.co.uk/news/world-africa-41861552.
37 Paul Harvey, 'Evidence on Corruption and Humanitarian Aid', CHS Alliance, 2 December 2015, available at https://www.chsalliance.org/news/blog/evidence-on-corruption-and-humanitarian-aid.
38 A.J. Duncan (n.d.), 'Is the Use of the Military in Complex Humanitarian Aid Operations a Political Quick Fix or Can it be the Cornerstone that Leads to Long Term Solutions?', Royal Logistic Corp Research Paper, 9.

'several NGOs struck me as being very narrowly focused',[39] while, in the case of Sierra Leone in 2014, British personnel reflected a view that NGOs dedicated a great deal of time and effort to negotiating the terms of their visual support at the expense of thousands of Ebola victims.[40]

Ultimately, the Ebola crisis illustrated a wide crisis of commitment and leadership, including within the WHO. Reports presented by *Newsweek* magazine revealed that the WHO and the World Bank appeared more concerned with the adverse effect a state of emergency would have on the region's markets, rather than focusing on containing the virus per se. In a study conducted by Laurie Garrett, it was recommended that future WHO responses should require 'a clearer mandate, better funding, more competent staff and less politicization'.[41] In response to such criticism, Margaret Harris of the WHO stated that 'It would have been much easier to control ... had we known [earlier] when the very first case occurred'.[42] With the benefit of hindsight, the international community missed a valuable window of opportunity to stem the spread of the contamination and, while their attention could and should have been focused consistently on the human risk, the ultimate response was undermined by a number of competing factors, including economic considerations, corruption, lack of knowledge of events on the ground and NGO aggrandisement, which are not within the control of any state or international organisation. As with Rwanda, the international community were more clearly motivated to deliver an effective response *after* the issue had received global attention, when wider international and national interests came into play. By this time, however, the challenges had increased significantly and a credible solution was much more difficult to deliver.

Case Study 3: Myanmar

Despite the UN declaration after the Holocaust that genocide would never again be tolerated by the international community – as enshrined in the 1948 Genocide Convention – recent events in Myanmar suggest that such

39 Gen Tim Cross, 'Comfortable with Chaos: Working with UNHCR and the NGO's, Reflections from the 1990 Kosovo Refugee Crisis', UN High Commission for Refugees Working Paper 42 (2001), 15.
40 Author's interview, 2018
41 Garrett, 'Ebola's Lessons', 102.
42 Westcott, 'Doctors Without Borders Slams Slow International Response to Ebola'.

declarations are of lesser value than the ringing rhetoric would imply. The principles of the 2005 Responsibility to Protect (R2P) were intended to provide a global safeguard and ensure that, in cases of humanitarian crisis or the violation of human rights, the international community would possess a framework and mechanism through which to take appropriate action.[43] Since its inception, however, the humanitarian philosophy motivating the originators of the R2P concept has been effectively undermined by structural selectivity in the UNSC: an international comfort blanket that has actually stifled progress towards more comprehensive action by frequently holding humanitarian impulses hostage to the ever-more partisan political vagaries of its permanent members and their power of veto. Nowhere is the initially promising, but now illusory, proposition of increased humanitarian action to respond to damaging state fragility more underwhelming than in Myanmar, where systematic ethnic cleansing of the Muslim Rohingya people has gone virtually unchecked by the wider international community.

Since 2012, relations between the Rohingyas in Rakhine state and the predominantly Buddhist government in Myanmar have deteriorated to the extent that, in 2017, the UN High Commissioner for Human Rights (UNHCR), Zeid Ra'ad al-Hussein, described their forced displacement into neighbouring Bangladesh as 'a textbook example of ethnic cleansing'.[44] The civilian establishment under State Counsellor Aung San Suu Kyi has been slow and reluctant to address claims the military have deliberately targeted Muslim homes and sponsored local paramilitary forces to engage in systematic property destruction, murder, rape and forced removals. According to BBC TV's *Panorama* programme, the attacks on the Rohingyas in August 2017 represented the 'climax of years of persecution and the result of deliberate military preparations'.[45] Since 1962, successive Burmese governments had stripped the Rohingya of their political and civil rights to the extent that they have been living in conditions some have equated to apartheid South Africa.[46] As a form of perverse explanation, the military

---

43 See *The Responsibility to Protect: Report of the International Commission on Intervention and State Sovereignty* (Ottawa: International Development Research Centre, 2001).

44 'UN human rights chief points to 'textbook example of ethnic cleansing' in Myanmar', available at https://news.un.org/en/story/2017/09/564622-un-human-rights-chief-points-textbook-example-ethnic-cleansing-myanmar.

45 'Myanmar: The Hidden Truth', Panorama, BBC1 12 December 2017.

46 Amnesty International 2018, 'Myanmar: New evidence reveals Rohingya armed group massacred scores in Rakhine State', available at https://www.amnesty.org/en/latest/news/2018/05/myanmar-new-evidence-reveals-rohingya-armed-group-massacred-scores-in-rakhine-state/.

establishment that still dominates the nominally civilian-led government has posited that the Rohingyas have never been a recognised community in Burma/Myanmar, and the issue of their presence has been ominously described by elements within the armed forces as an 'unfinished problem'.[47] There is clear evidence the military have – and are likely to continue to pursue – a strategy of 'clearing' the Rohingya along the Bangladeshi border, safe in the knowledge they will not be held to account by the international community. This should not be a surprising assumption given the longer-term historical developments in this case and the other examples of international passivity noted in this chapter.

To date, the response of the international community to this case has been primarily limited to public denunciations of Myanmar's military, restrictions on weapons sales, the banning of the training of Myanmar's officers in the West (which, under Pillar Two of R2P, could have been considered part of an 'assistance' programme to create better, more credible and more humane indigenous capacity) and, in a unique move, a boycott of Myanmar's ruby market. Such steps are unlikely to be sufficient to positively change the situation for the Rohingyas and they could be interpreted as the West offering little more than lip service once again to the 'humanitarian' concerns that seem to so often follow in the wake of increased state fragility, arguably as both cause and consequence. It also suggests the familiar refrain of the international community – particularly in its unrepresentative form within the UNSC – seeking to prioritise other geopolitical and geostrategic challenges over the humanitarian needs of the Rohingya. Whatever the West's intentions towards Myanmar (if any), the outflow of half a million refugees has remained a local issue. In another eerily depressing parallel to Rwanda, Myanmar's limited strategic value to the West suggests it is likely the Rohingya will be left largely to their own fate. Unless a security reassessment highlights Myanmar's instability as a clear global risk for the West, the situation there is likely to be added to the already lengthy list of ineffective humanitarian interventions in fragile states.

Conclusion

It is inevitable that international scholars, politicians and the global military community will continue to face challenges when confronted with

---

47 'Myanmar: The Hidden Truth'.

the humanitarian and strategic complexities arising from state fragility. At this point, the world is largely devoid of strategic theorists who deal exclusively or predominantly with humanitarian relief operations and the stability they might offer. Perhaps, instead of restricting themselves to selectively launching ad hoc humanitarian/state-building initiatives, of varied commitment in terms of timing, resources, manpower and focus, the international community could commit to developing a broader comprehensive doctrine than that of the equally inconsistent R2P: one that takes account of past experience and is more cognisant of the demands of those most directly affected by such problems.

This chapter has illustrated – through a number of key historical and contemporary case studies – both the range of challenges the international community continues to confront when faced with such deteriorating situations and the generally ineffective responses, both of the international community at large and the West in particular, while accepting the precarious nature of international legal dispensation through the unreformable UN veto structure serves as both stumbling block and convenient scapegoat to evade responsibility to act. It is unsurprising, given this record, that there is limited confidence in the international community to live up to its own rhetoric on such matters. Yet, as recent events in West Africa and the exposure of exploitation and corruption among NGOs' global operations has shown, a number of key, non-state actors also demonstrate significant flaws.

Only by constructing an agreed operating framework for international organisations, governments and NGOs – one that includes developing and fragile states themselves – can a set of comprehensive-yet-agile intervention options be agreed upon and developed. Such a framework could offer a mechanism at the very least to help manage expectations better. Organisations could be held to account across a broad range of operating factors, from the time to deploy credible forces to realising measurable end states. At the very least, the framework could identify the range of disciplines the international community needs to maintain if credible response capabilities are to be held as contingency. Without such a framework, the Western-dominated international organisations are consigned to reacting to vulnerable situations where national interests are allowed to dictate the form of each response, leading to highly variable outcomes. Until then, the realist basis for understanding international relations prevails: we are all, ultimately, on our own.

References

African Rights, *Rwanda: Death, Despair and Defiance*. London: African Rights, 1995.

Barnett, M. *Eyewitness to Genocide: The United Nations and Rwanda*. Ithaca NY: Cornell University Press, 2002.

Bellamy, A.J. and N.J. Wheeler. 'Humanitarian Intervention in World Politics', in J. Baylis, S. Smith and P. Owens (eds). *The Globalization of World Politics: An Introduction to International Relations*. Oxford: Oxford University Press, 2011.

Brock, L. et al. *Fragile States: Violence and the Failure of Intervention*. Cambridge: Polity Press, 2012.

Clarke, W. and J. Herbst. 'Somalia and the Future of Humanitarian Intervention', *Foreign Affairs* 75:2, 1996.

Cross, T. 'Comfortable with Chaos: Working with UNHCR and the NGOs, Reflections from the 1990 Kosovo Refugee Crisis', UN High Commission for Refugees, *Working Paper* 42, 2001.

Dallaire, R. *Shake Hands with the Devil: The Failure of Humanity in Rwanda*. Toronto: Vintage Canada, 2004.

Falk, R. *Achieving Human Rights*. London: Routledge, 2009.

Garrett, L. 'Ebola's Lessons: How the WHO Mishandled the Crisis', *Foreign Affairs* 94:5, 2015.

Harvey, P. 'Evidence on Corruption and Humanitarian Aid'. CHS Alliance, 2 December 2015.

ICISS, *The Responsibility to Protect: Report of the International Commission on Intervention and State Sovereignty*. Ottawa: International Development Research Centre, 2001.

Kiernan, B. *Blood and Soil: A World History of Genocide and Extermination from Sparta to Darfur*. London: Yale University Press, 2007.

McMahon, S. 'Op GRITROCK – Fighting Ebola and the Legacy of Defence Engagement'.,*British Army Review* 164, 2015.

Melvern, L.R. *A People Betrayed: The Role of the West in Rwanda's Genocide*. London: Zed Books, 2006.

Vilmer, J.B.J. 'Does the 'CNN Effect' exist? Military Intervention and the Media'. IGN Global, 28 June 2012.

Walsh, S. and O. Johnson. *Getting to Zero: A Doctor and a Diplomat on the Ebola Frontline*. London: Zed Books, 2008.

Wheeler, N.J. *Saving Strangers: Humanitarian Intervention in International Society*. Oxford: Oxford University Press, 2000.

4

SOMALIA AND ITS NEIGHBOURS

AMISOM and the Regional Construction of a Failed State

Jonathan Fisher[1]

Introduction

This chapter focuses on explaining regional state actors' understandings of regional state-building in Somalia through, and in the context of, the African Union Mission in Somalia (AMISOM) operation. It looks primarily, therefore, at the period since the mission's genesis in 2005.[2] It focuses

1   A version of this chapter was published as an article ('AMISOM and the regional construction of a failed state in Somalia') in *African Affairs* in September 2018, and the author is grateful to the editors of that journal for allowing the reproduction of parts of it here. The article derived, in part, from a collaborative project undertaken with Sally Healy, and the author is grateful to Sally for her advice and feedback during the early development of the study. He would also like to thank David Brown, Nic Cheeseman, Nicolas Lemay-Hébert, Martin Smith and the editors and anonymous reviewers of *African Affairs* for valuable feedback on earlier versions of this chapter. He also acknowledges support from the Economic and Social Research Council (Grant Number ES/N008367/1), which funded a seminar in Nairobi in September 2017 where some of the ideas in this chapter were presented and developed. He is grateful, finally, to participants in the symposium held at Sandhurst Military Academy on 'Fragile States: Challenges and Responses' in February 2018 for comments on an earlier draft.
2   Jonathan Fisher, 'Managing donor perceptions: Contextualizing Uganda's 2007 intervention in Somalia', *African Affairs* 111:444 (2012), 421–2.

particularly on the years 2007–14, when most states in the region came to transform their formerly semi-independent approaches to state-building in Somalia into formal membership of AMISOM; Uganda joined the mission in 2007, Djibouti in 2011, Kenya in 2012 and Ethiopia in 2013. The broader history of regional engagement with, and involvement in, Somalia has nevertheless long informed the thinking and behaviour of many regional players and is discussed, briefly, in the second half of the chapter. Methodologically, the study is principally concerned with uncovering how regional AMISOM troop-contributing state actors (specifically Djibouti, Ethiopia, Kenya and Uganda)3 have conceived of the state-building project in Somalia – what they have considered to be a desirable form of political authority in the country, what they believe their role should be in fostering such authority and what ideas and factors have lain behind these perspectives.

Evidence used to answer the chapter's core questions is identified through examining regional actors' articulations of their views of, and visions for, Somalia. Understanding the character of international relations to be socially constructed, the study places emphasis on how ideas on state failure and the role of regional actors shape, determine and, often, are used to justify actions. This is a common approach taken in scholarly critiques of international interventions but not, to date, of regionally led operations, for reasons which remain unclear. The chapter does not claim to provide a comprehensive exploration of the perspectives of each of the four regional AMISOM contributors considered and, instead, focus is placed on examining politico-military elites from these states as an 'epistemic community', at least in relation to state-building in Somalia.[4]

The chapter analyses actor statements and narratives produced in contexts where key regional exchanges on Somalia have taken place and where regional perspectives are negotiated and consolidated. Particularly central are transcripts of high-level regional meetings and summits held between 2004–11 under the aegis of the Intergovernmental Authority on Development (IGAD) and accessed during a visit to IGAD's paper archives during May 2014. Fifteen documents relating to Somalia were reviewed,

---

3   Burundi is not included given its limited involvement in the regional politics of the Greater Horn of Africa outside AMISOM.

4   Haas describes an epistemic community as 'a network of professionals with recognised expertise and competence in a particular domain and an authoritative claim to policy-relevant knowledge within that domain'. See Peter M. Haas, 'Introduction: Epistemic communities and international policy coordination', *International Organization* 46:1 (1992), 1–35.

with five directly cited below. In addition, 17 author interviews with IGAD officials and Djiboutian, Ethiopian, Kenyan and Ugandan officials are drawn upon. These interviews have been undertaken across a range of fieldwork trips to Djibouti, Ethiopia, Kenya and Uganda between 2013 and 2017. In most cases, interviewees asked that they not be identified by name or office, and only a selection of interviews are cited. Finally, the analysis and argument is also informed by the author's participation in two regional dialogue fora hosted by the Friedrich Ebert Stiftung in Nairobi during 2014–15.[5]

'Failed States', Regional Intervention and Somalia

The 'failed state' concept entered scholarly and policy discourses during the mid-1990s and has since become a widely acknowledged – albeit contested – 'category' of state for many academics and practitioners.[6] Qualified by, and in competition with, a range of other terms, the 'failed state' is generally agreed by scholars, who favour the term to be the end-point at which political, social, economic and security institutions disintegrate, sometimes simultaneously.[7]

Contributors to this literature have focused not only on 'diagnosing' state failure but also on prescribing cures, even adopting, in some cases, in the words of Nicolas Lemay-Hébert, 'diagnostic medical analog [ies] to exemplify how we should be able to forecast state failure'.[8] This reflects not only the policy focus of much of this school of thought, but also its conviction that functioning political authorities possess core characteristics that can – and should – be replicated to restore order and predictability in conflict and post-conflict situations.

5   The Friedrich Ebert Stiftung is a German political foundation. Its annual Greater Horn of Africa Dialogue brings together regional and international political, military and civil society stakeholders as a means to 'facilitate political dialogue on [regional] security threats and ... national, regional and continental responses'. See Antonia Witt, *10th FES annual conference: Peace and Security in the Horn of Africa* (Addis Ababa: Friedrich Ebert Stiftung, 2014), 5.

6   Sonja Grimm, Nicolas Lemay-Hébert and Olivier Nay, '"Fragile states": Introducing a political concept', *Third World Quarterly* 35:2 (2014), 197–209.

7   Robert Rotberg, *When States Fail: Causes and Consequences* (Princeton NJ: Princeton University Press, 2003), 1–25.

8   Nicolas Lemay-Hébert, 'Rethinking Weberian approaches to statebuilding', in David Chandler and Timothy Sisk (eds), *Routledge Handbook of International Statebuilding* (London: Routledge, 2013), 6. Prominent examples include Ashraf Ghani and Clare Lockhart, *Fixing Failed States: A Framework for Rebuilding a Fractured World* (Oxford: Oxford University Press, 2008); Gerald Helman and Steve Ratner, 'Saving failed states', *Foreign Policy* 89 (1992–3) and Rotberg, *When States Fail*.

This literature draws heavily on the writings of German sociologist Max Weber (1864–1920) and his characterisation of the state as a 'compulsory political association … [whose] administrative staff successfully upholds the claim to the monopoly of the legitimate use of physical force in the enforcement of its order'.[9] This approach to understanding political authority has focused on ascribing stability and statehood to formal institutional bodies that maintain a 'monopoly of violence' across a particular territory, through armed forces of various kinds but also governance mechanisms, including judicial and bureaucratic systems.[10]

According to this analysis, the state is composed of legal-rational office-holders and the bulk of United Nations (UN) and other peacekeeping missions since the 1990s have been informed by a similar logic: that functioning states require particular governance structures and machineries, whose basic shape and logic can be effectively replicated across different contexts.[11]

Many of the most expansive of these missions came into being in the years following the Cold War, when notions of global democratic victory persuaded designers of peace operations that successful post-conflict state-building could be secured through transplanting institutions focused around a liberal democratic political order and a liberal market-based economic order, seemingly without regard for existing systems of political authority in intervention sites.[12] African states have featured prominently in academic and policy discussions of state failure and liberal interventionism. Somalia, in particular, invariably appears near the top of state failure indices and has long been regarded by 'problem solver' commentators, such as Robert Rotberg and Eben Kaplan, as the 'model of a collapsed state … the classical failed … state',[13] and the 'very definition of a failed state'.[14]

Critically for this study, however, such perspectives have often been internalised and developed by African political and intellectual elites

9   Max Weber, *Economy and Society, Part 1 and 2*, trans Günther Roth and Claus Wittich (Berkeley CA: University of California Press, 1978), 54.

10 Joel Migdal, *Strong societies and weak states: State-society Relations and State Capabilities in the Third World* (Princeton NJ: Princeton University Press, 1988); Charles Tilly, *Coercion, Capital and European States, AD 990–1990* (Oxford: Blackwell, 1990).

11 Roland Paris, *At War's End: Building Peace After Civil Conflict* (Cambridge: Cambridge University Press, 2004).

12 Oliver Richmond, 'The problem of peace: Understanding the 'liberal peace'', *Conflict, Security and Development* 6:3 (2006), 291–314.

13 Robert Rotberg, 'Failed states in a world of terror', *Foreign Affairs* 81:4 (2002), 127–40.

14 Eben Kaplan, *Somalia's terrorist infestation* (New York: Council on Foreign Relations, 2006); Rotberg, *When States Fail*, 11–12.

themselves. In 1995, influential Kenyan scholar Ali Mazrui delineated six 'basic functions of the state' in an article on 'the failed state' in Africa, drawing heavily upon neo-Weberian frameworks to argue that 'it is clear that many [African] states are in trouble'.[15] More recently, the African Development Bank's Fragile States Facility (now Transition Support Facility) has produced a framing document that defers to how 'most development agencies broadly describe fragile states' – a description which focuses on 'state institutions' – in their explorations of the concept.[16]

It is important to recognise the significance of the Global War on Terror (GWOT) for understanding how neo-Weberian portrayals of state integrity have gained salience in East Africa. Many of the region's key aid donors – notably the US, UK and European Union (EU) – have calibrated their relationships with East African states around countering Islamic extremism and denying 'safe haven' to Islamist terrorists since 9/11 especially.[17] This has compelled the region's politico-military networks, many of whose operations depend on Western financial and military support, to engage with underlying donor policy assumptions around state failure and the place of political Islam within regional politics. This has not been a unidirectional process by any means – the author has previously written on African states' framing of, for example, concepts of 'failed' statehood in international fora[18] – but has nonetheless provided a critical regional meeting place for debates on, and articulations of, state fragility and the shape of 'acceptable' political authority. This has had important implications for regional approaches to state-building in Somalia, as detailed below.

These neo-Weberian perspectives have been assailed from multiple directions. Africanist scholars have long questioned even the *ability* of some African states to project power beyond urban centres.[19] More recent discussions on 'hinterland state failure' and 'security pluralism' take this debate further by highlighting when African states *choose* not to project or sustain state control in and over certain parts of their claimed territories

---

15 Ali Mazrui, 'The blood of experience: The failed state and political collapse in Africa', *World Policy Journal* 12 (1995), 28.

16 Mthuli Ncube and Basil Jones, *Drivers and dynamics of fragility in Africa*, Chief Economist Complex, Africa Economic Brief 4, 5 (Abidjan: African Development Bank, 2013), 1.

17 Jonathan Fisher and David M Anderson, 'Authoritarianism and the securitisation of development in Africa', *International Affairs* 91:1 (2015), 131–51. See also the discussions by Emily Knowles and Abigail Watson in chapter 8 of this volume.

18 Jonathan Fisher, "'When it pays to be a fragile state': Uganda's use and abuse of a dubious concept', *Third World Quarterly* 35:2 (2014), 316–32.

19 Jeffrey Herbst, *States and Power in Africa: Comparative Lessons in Authority and Control* (Princeton NJ: Princeton University Press, 2000).

for reasons of pragmatism, strategy and regime maintenance.[20] External impositions of one-size-fits-all state architectures have also been criticised as inappropriate, including in studies of Somalia, where alternative forms of political authority, such as Somalia's Islamic Courts Union (ICU),[21] have been recognised, or where localised manifestations of political order have been contrasted with poorly functioning but internationally recognised state institutions.[22]

To date both sides of the debate have tended towards contrasting 'international' practices and perceptions with those of the 'local'.[23] Regional missions, such as AMISOM, have often been conceptualised as proxies for the international system in this regard, with the distinctive role and place of the region being overlooked. This study locates itself within this conceptual gap, focusing on how regions – as opposed to international organisations/Western donor countries or local communities – construct and conceptualise state failure and how best to respond to it. Since the establishment of the African Union (AU) in 2002, peacekeeping and peacebuilding have become an increasingly regionalised affair in Africa, even if operations have continued to be funded largely by states outside the continent. The AU has established peacekeeping missions in Burundi, Sudan, Darfur, Somalia, Mali and the Central African Republic since 2003, while regional bodies formerly with little involvement in security affairs have created standby forces and brigades to respond to regional crises.[24] Most of these missions are led by neighbours of the state being intervened in and most have been justified by the intervening states – and Western aid donors – in terms of providing 'African solutions to African problems', a concept first developed by US officials as a means to justify withdrawal of

---

20 Ken Menkhaus, 'State failure and ungoverned space', in *Ending Wars, Consolidating Peace: Economic Perspectives* (London: International Institute for Strategic Studies, 2008), 171–88; Mareike Schomerus and Lotje de Vries, 'Improvising border security: "A situation of security pluralism" along South Sudan's borders', *Security Dialogue* 45:3 (2014), 279–94.

21 An association of Sharia courts which effectively controlled much of south/central Somalia between June–December 2006.

22 Tobias Hagmann and Markus Hoehne, 'Failures of the state failure debate: Evidence from the Somali territories', *Journal of International Development* 21:1 (2009), 42–57; Ken Menkhaus, 'State failure, state-building, and prospects for a 'functional failed state' in Somalia', *The Annals of the American Academy of Political and Social Science* 656:1 (2014), 154–72.

23 Severine Autesserre, *The Trouble with Congo: Local Violence and the Failure of International Peacebuilding* (Cambridge: Cambridge University Press, 2010); Ghani and Lockhart, *Fixing Failed States*.

24 Paul D. Williams and Arthur Boutellis, 'Partnership peacekeeping: Challenges and opportunities in the United Nations–African Union relationship', *African Affairs* 113:451 (2014), 254–78.

American troops from regional conflict theatres, but which has since been appropriated and instrumentalised by African political elites as a rallying cry for African unity, agency and self-help.[25]

Understanding the contemporary dynamics of state-building, in/security and international relations in African conflict spaces therefore necessitates a reconsideration of the role of regional states, particularly through the mechanism of regional peacekeeping, and a focus on what is often an analytical 'missing middle' between international and (notionally) African or local perspectives. The remainder of this chapter will provide this focus through exploring regional approaches to state-building in Somalia, via AMISOM, following a brief overview of the mission's provenance, evolution and context.

AMISOM: The Regionalisation of the Somalia Conflict

The regime of Somalia's Siad Barre – in power since October 1969 – collapsed in January 1991. The steady removal of political and economic opportunities from stakeholders and clans outside the president's own clan base during the 1980s is widely regarded as being the key mobiliser of multiple rebellions that eventually led to the longstanding Somali leader's downfall. This situation was exacerbated by a sudden loss of international support for Barre during the initial post-Cold War years, as Somalia's previous strategic value rapidly evaporated.[26] Since this time, Somalia has been governed by multiple political authorities. While the northern territories of Somaliland and Puntland – which *de facto* seceded from the rest of the country in 1991 and 1998 respectively – have experienced a re-establishment and consolidation of some measure of political order, the rest of Somalia, often referred to as 'south/central Somalia' and the focus of AMISOM and this chapter, has not.

AMISOM – which worked in tandem with Ethiopian troops until the latter's amalgamation into the operation in 2014 – has operated with considerable success in urban areas. It has not been the ICU that AMISOM and its allies have fought against for most of this time, but a far more militant off-shoot of the Union, Al-Shabaab. Outside of towns, Al-Shabaab remains

---

25 Ricardo Soares de Oliveira and Harry Verhoeven, 'Taming intervention: Sovereignty, statehood and political order in Africa', *Survival* 60:2 (2018), 7–32.
26 Abdi Ismail Samatar, 'Destruction of state and society in Somalia: Beyond the tribal convention', *Journal of Modern African Studies* 30:4 (1992), 625–41; I.M. Lewis, *Understanding Somalia and Somaliland: Culture, History and Society* (London: Hurst, 2008).

AMISOM and the Somali Federal Government (SFG)'s primary adversary and, indeed, the major power in many rural areas of southern Somalia, their influence and ability to mobilise buoyed by AMISOM's unpopularity.[27] The mission's heavy focus on counter-insurgency for much of its lifetime has led leading specialists – such as Paul D. Williams – to conceptualise it as a counter-insurgency or, at best, a *peace* operation rather than a peace*keeping* operation.[28]

For Djibouti and Kenya, the facilitation of peace conferences had been the preferred vehicle for dealing with the Somali crisis until the later 2000s.[29] Investing significant diplomatic capital and resources in these processes, Kenya facilitated the election and emergence of the Somali transitional government (the Transitional Federal Government or TFG) and parliament in 2004 and hosted the TFG and its institutions until their 2006 move to Somalia itself. Between 2008 and 2009, Djibouti hosted UN-led negotiations on the evolving composition of the TFG leadership. Nairobi continues to host those diplomatic missions and aid agencies accredited to but unable, or unwilling, to relocate permanently to Mogadishu on the basis of agency assessments of potential risks to personnel deployed to the Somali capital.

The rise of Al-Shabaab during the later 2000s reinvigorated longstanding concerns in Nairobi regarding the security threat posed to Kenya by Islamists in Somalia, particularly following kidnapping incidents in northern Kenya in autumn 2011. In October 2011, Kenya took the unprecedented step of intervening militarily in Somalia in an effort to crush Al-Shabaab, and Kenyan troops remain today, albeit 're-hatted' as part of AMISOM in 2012.[30] Where Ugandan AMISOM troops have focused primarily on shoring-up the TFG (and later SFG) in Mogadishu, Kenyan forces have instead remained around the border region between Kenya and Somalia, where they have played a significant role in shaping local politics. Djibouti also joined the mission in 2011.

In recent years, while Somalia has remained a hotspot of conflicting regional agendas and interests, regional actors have aligned more around

---

27 For more on Al-Shabaab and the local dynamics of AMISOM's (un) popularity, see David M. Anderson and Jacob McKnight, 'Kenya at war: Al-Shabaab and its enemies in Eastern Africa', *African Affairs*, 114:454 (2015), 1–27; Stig Jaarle Hansen, *Al-Shabaab in Somalia: The History and Ideology of a Militant Islamist Group, 2005–2012* (Oxford: Oxford University Press, 2012); Deborah Valentina Melito, 'Building terror while fighting enemies: How the Global War on Terror deepened the crisis in Somalia', *Third World Quarterly* 36:10 (2015), 1–21.
28 Walter Lotze and Paul D. Williams, *The Surge to Stabilize: Lessons for the UN from the AU's Experience in Somalia* (New York: International Peace Institute, 2016).
29 Kidist Mulugeta, *The role of regional and international organizations in resolving the Somali conflict: The case of IGAD* (Addis Ababa: Friedrich Ebert Stiftung, 2009).
30 Anderson and McKnight, 'Kenya at war'.

a single instrument to reconstruct the Somali state than at any time since the fall of Siad Barre. Through AMISOM, all of the country's immediate neighbours, together with Uganda, Burundi and Sierra Leone, have troops on the ground and all remain theoretically committed to strengthening the SFG polity, it being the permanent successor of the TFG, inaugurated in 2012. It was not, however, inevitable that regional actors would coalesce around such an approach, nor that they would identify their role as being to support and undergird a set of formal political institutions devised outside Somalia. To explain how this coming together of regional opinion occurred, the remainder of this study explains how regional states came to discursively construct Somalia as a 'failed state' requiring outside intervention and the establishment of a particular kind of neo-Weberian political order.

AMISOM and the Region I: Constructing a 'Weak, Fragile and Failed State'

East African political leaders have, understandably, often forcefully rejected the imposition of 'state fragility' labels onto their own polities by international actors and institutions. They have not, however, balked at applying such labels to Somalia themselves in recent years. Ugandan officials have been particularly prominent in this regard. In seeking to persuade parliamentarians to vote in support of Uganda contributing troops to AMISOM in February 2007, Defence Minister Crispus Kiyonga argued that 'Somalia has been a failed sister African state for nearly 16 years'.[31] Later that year, Foreign Minister Sam Kuteesa privately justified Uganda's involvement in Somalia to Eritrean President Isaias Afwerki by arguing that 'Uganda [like Somalia] was once a failed state that needed its neighbours' help'.[32] In a March 2008 conference of regional ministers in Kampala, Ugandan President Museveni opened the meeting by highlighting the importance of joint operations in states such as Somalia as a means to ward off 'the dangers of weak, fragile and failed states' while, in September 2010, Kuteesa told a Kampala newspaper that Ugandan intervention in Somalia had been premised on 'mak [ing] sure there is no failed state [in the region] '.[33]

---

31 Parliament of Uganda proceedings, *Hansard*, 13 February 2007.
32 US Embassy Kampala diplomatic cable leaked by *Wikileaks*, 'Uganda: Official discussions with Eritrea on Somalia', 20 April 2007, available at https://wikileaks.org/plusd/cables/07KAMPALA669_a.html.
33 Report of the 3rd extra-ordinary meeting of defence and security ministers of IGAD, Kampala, 12 March 2008 (IGAD Secretariat Archives, 12 May 2014); Shifa Mwesigye, 'INTERVIEW: People will get rid of Museveni – Kutesa', *Observer* (Kampala), 1 September 2010.

Somalia's immediate neighbours have been less prepared to present Somalia as 'failed' *per se*, but they have frequently characterised the polity and their involvement therein using similar language and sentiments. In May 2013, at an international summit on Somalia attended by regional counterparts, Djiboutian President Ismaïl Guelleh characterised Somalia's recent past as a period of 'conflict, political dissensions, warlordism, extremism, piracy'.[34] A year later, this was echoed by a Djiboutian official based at IGAD, who argued that 'this place Somalia, it has been a place of chaos for so long ... it is somewhere there has been no government'.[35]

In Kenya, then-President Mwai Kibaki highlighted the 'lawlessness' in contemporary Somalia to attendees at a 2011 state dinner and the 'fragile' situation in the country before assembled counterparts at the UN in 2008,[36] while Kenyan Deputy President William Ruto lamented to viewers of Kenya's *Citizen TV* in July 2014 that the country had been 'babysitting the situation in Somalia for 30 years'.[37] Kibaki's successor, Uhuru Kenyatta, told CNN in October 2015 that 'you know, we had a failed state right next to our border, a state where there was no rule of law, there was no government, and it was just open vast land'.[38] This followed a 2014 editorial in Kenya's *Daily Nation*, written by Kenyatta, which highlighted the threat to Kenya from Somalia's purportedly ungoverned spaces.[39] In Ethiopia, Foreign Minister Seyoum Mesfin noted in a 2007 media interview that 'Somalia was a failed state for the last 15–16 years ... Somalia is a country that has failed'.[40] A decade later, one of his chief aides discussed with this author Ethiopia's historical attempt to 'come up with a cure' for Somalia's 'failure to function'.[41]

The argument here is not that regional characterisations of the Somali polity are normatively problematic or inaccurate as such. Nor are they necessarily a significant departure from characterisations promoted by

---

34 Speech by Ismaïl Guelleh, President of Djibouti, Somalia Conference, London, 7 May 2013, available at https://www.youtube.com/watch?v=tt3750-Zn8A/.

35 Author's interview with senior IGAD official, Djibouti May 2014.

36 Kevin Kelley, 'Country back on track, Kibaki tells UN', *Daily Nation* (Nairobi), 24 September 2008; 'Kibaki: Somalia instability a concern', *Daily Nation*, 1 April 2011.

37 *Citizen TV*, 'Big Question', 1 July 2014.

38 *CNN Press Room*, 'President Kenyatta on gay rights in Kenya', Full interview transcript, 18 October 2015, available at http://cnnpressroom.blogs.cnn.com/2015/10/18/president-kenyatta-on-gay-rights-in-kenya/.

39 Uhuru Kenyatta, 'Our country and our people are under attack; it is a war we must win together', *Daily Nation* (Nairobi), 2 December 2014.

40 Robert Wiren, 'Interview: Seyoum Mesfin', *Les nouvelles d'Addis*, 12 December 2007, available at www.ethiomedia.com/access/interview_seyoum_mesfin.html.

41 Author's interview with senior Ethiopian foreign ministry official, Addis Ababa September 2017.

northern states and development agencies. It is not surprising that the leaders of states whose citizens have died in terrorist attacks perpetrated by Al-Shabaab have framed Somalia in terms of instability, unpredictability and fragility. The point, rather, is that, in presenting their view of Somalia to domestic, regional and international audiences, East African governments have reified strong/fragile state dichotomies and discourses of diagnosing and measuring state failure. Critically, this portrayal of Somalia as failed has been interlaced with, and has undergirded, regional discourses justifying regional interference, both political and military. Scholars have critiqued the international community's use of such rhetorical sleights of hand but not, to date, that of *regional* actors.[42]

One central discourse in this regard has focused around the threat posed to neighbours by a lawless Somalia. One senior figure from a regional state argued to counterparts in a roundtable discussion in Nairobi in 2015 that 'if Somalia is not at peace, Kenya cannot be at peace, Sudan cannot, Ethiopia cannot', while a senior IGAD Peace and Security official noted in a May 2014 interview that 'conflict in the region is intertwined and goes beyond national frontiers; with the Somali conflict the peace of Ethiopia, Kenya and Djibouti is affected'.[43]

This notion of Somali insecurity as a regional cause for concern has been incorporated into a wider East African elite narrative around regional obligations and, indeed, as an instrument to bring regional actors together around a common foreign policy initiative. In a 2004 meeting of East African ministers in Kigali, IGAD Executive-Secretary Attalla Bashir told assembled figures that 'the region should see [the] Somalia [crisis] as a unique opportunity to play a pivotal role ... to create strong and viable security structures in the region'.[44] In October 2008 his successor, Mahboub Maalim, told regional foreign ministers that 'full deployment of AMISOM is a condition that is absolutely necessary to save Somalia'.[45] Previously, in March 2006, Museveni had chided regional counterparts for 'a lack of cooperation among the states of the region', which 'turns small problems

---

42 Menkhaus, 'State failure'.
43 Comments by senior figure from an East African state, Horn of Africa Dialogue Forum, Nairobi, 3 November 2015; Author's interview with senior IGAD Peace and Security official, Djibouti, 13 May 2014.
44 Report of meeting of East African ministers of defence and security, Kigali, 9–10 September 2004 (IGAD Secretariat Archives, 13 May 2014).
45 Report of extraordinary meeting of the IGAD Council of Ministers, Nairobi, 28 October 2010.

into big problems'.[46] These sentiments have often been framed in terms of a moral obligation to intervene following the 1994 Rwandan genocide, but have rarely been accompanied by reflection on, or consideration of, the agency or opinions of Somali actors themselves. Indeed, in accepting and internalising the 'failed state' narrative, regional actors have positioned themselves beyond a point where accessing such opinions might be necessary.

Regional characterisations of Somali state integrity – and regional obligations therein – have rested to a considerable degree on internalised neo-Weberian institutionalist frameworks and discourses. The development of this regional consensus around militarily resolving the Somali 'problem', however, also rests on a second, more longstanding regional portrayal of the Somali polity – and Somalis themselves – as a threat to the postcolonial regional order that requires containment. The Somali people of the Horn share a common language, culture, religion and adherence to a common system of customary law, but they were divided after the 1880s between five separate administrations, including three European colonial powers. In the complex, protracted and multi-layered negotiations leading to the creation of contemporary Ethiopia, Kenya and Somalia, however, the cause of a unified Somali nation was ultimately subordinated to a range of other concerns and interests.[47]

This existential mismatch between Somali nationhood and statehood has presented a challenge to Nairobi and Addis Ababa in particular, since successive Somali administrations have supported irredentist Somali insurgencies in their borderlands, often resulting in major crackdowns.[48] Somalia's 1977 invasion of Ethiopia's Ogaden region to 'reclaim' Somali-inhabited territories helped cement, in the words of one longstanding Ethiopian foreign ministry official, the notion that 'Somalia was the enemy, Somalia has always wanted to undermine Ethiopia'.[49] Regional constructions of Somalia's 'lawlessness' spilling over borders therefore are based not only in internalised neo-Weberian models, but also on a more enduring regional portrayal of Somalia as an existential threat to established postcolonial statehood. The same is true for the development of a regional consensus around intervening in Somalia.

---

46 Proceedings of the 11th summit of the assembly of heads of state and government of IGAD, Nairobi, 20 March 2006 (IGAD Secretariat Archives, 12 May 2014).
47 Daniel Branch, 'Violence, decolonization and the Cold War in Kenya's north-eastern province, 1963–1978', *Journal of Eastern African Studies* 8:4 (2014), 642–57.
48 David M. Anderson, 'Remembering Wagalla: State violence in northern Kenya, 1962–1991', *Journal of Eastern African Studies* 8:4 (2014), 658–76.
49 Author's interview with former senior Ethiopian foreign ministry official, Addis Ababa, 14 September 2017.

AMISOM and the Region II: Constructing 'the Very Essence of Statehood'

In delineating the kind of state they wish to see established, or reconstituted, in Somalia via AMISOM, regional actors have also demonstrated limited interest in building on existing forms of authority or on local Somali preferences. Instead, the imposition of a particular idea of what a state is has been a core concern – this idea largely resembling the model of statehood outlined by neo-Weberian theorists above. At the heart of the AMISOM conception of the Somali state has been the army and security services – convened, assembled and trained by regional forces, often outside Somalia. Ethiopian and Ugandan officials in particular have viewed the 'building of a Somali army' as central to the reconstituted Somali state, with Museveni describing a professionalised, national army as 'one of the most important pillars of the state' at a 2013 IGAD summit.[50] This focus also became one of the central issues exercising regional actors in advance of major international conferences in London in 2013 and 2017. The establishment and consolidation of new formal political institutions, namely an executive, a parliament, regional administrations and district administrations, have also been viewed as critical foundations of civil authority by regional powers.[51] In October 2008, Kenyan Foreign Minister Moses Wetangula summarised the region's perspective on what this state should look like before regional counterparts, arguing that 'we must devise an expert–informed collective approach to Somalia, otherwise it will be business as usual. We should refocus support to the Transitional Federal Institutions which are the very essence of statehood'.[52]

This focus on building formal institutions was further fleshed out by Ethiopian and Ugandan officials in interviews in 2013, when this author was informed that regional state-building priorities in Somalia were focused around 'building the capacity of the Somali security services, particularly command and control structures'.[53] These structures do not appear, however, to have been designed to complement or align with existing manifestations of political authority in Somalia or to contribute to the negotiation of a social

50 *New Vision* (Kampala), 'Professionalise Somali army – Museveni', 4 May 2013; Author interviews with UK, EU, Ethiopian, Kenyan and Ugandan officials, Addis Ababa, May 2013 and April 2016, and Nairobi, November 2014 and June 2016.

51 Fred Oluoch, 'Kenya is not acting at the behest of US!', *Daily Nation* (Nairobi), 9 January 2007.

52 Report of the 29th assembly of the council of ministers, Nairobi, 28 October 2008.

53 Author's interview with senior Ethiopian diplomat, Addis Ababa, May 2013.

contract to undergird the nascent state.[54] As Williams has noted, 'AMISOM… remains a predominantly military operation' with limited capacity or interest in moving from offence and defence to 'stabilization tasks'.[55] Indeed, for some regional officials interviewed for this study, there was a sense that regional actors understand better what kind of political setup Somalia requires than Somalis themselves. One regional diplomat posted to Addis Ababa noted, for example, that 'Somalis are still patients in an intensive care unit … you can't expect them to know about governance yet'.[56]

This imagining of statehood as, first and foremost, defined by the establishment of a strong, disciplined army is clearly based in neo-Weberian state-building frameworks. It also, however, constitutes neighbouring regimes' seamless transference of their own approaches to post-conflict reconstruction to the Somali context. The current Ethiopian and Ugandan governing elites, for example, emerged from guerrilla liberation movements, whose approach to state-building domestically has been founded around military logics and the armed forces as an ideological vanguard.[57] As in other post-liberation polities, this body is envisaged as a transformative institution tasked not only with defence and security, but also economic governance, norm development and service provision.[58] Neither Addis Ababa nor Kampala appear to have considered whether this model of military-first statehood is appropriate to the quite different Somali context, outlined below. One senior Ugandan and AU official noted in July 2017, for example, that Uganda 'had come with a medicine based on the experience of Uganda's liberation struggle, even if it is taking some time for the Somali people to take the medicine'.[59]

Indeed, regional state elites have tended to reject Somali political structures and norms where they do not conform to governance models elsewhere in the region. Alternative, context-specific forms of political authority – including those based around notions of heterarchy (a polity where multiple actors compete for, or exercise, political authority) – have

54 Alice Hills, 'Somalia works: Police development as state building', *African Affairs* 113:450 (2014), 88–107.

55 Paul D. Williams, 'After Westgate: Opportunities and challenges in the war against Al-Shabaab', *International Affairs* 90:4 (2014), 916–17.

56 Author's interview with senior Ugandan military official, Addis Ababa, May 2013.

57 Risdel Kasasira, 'Wikileaks: Museveni discredits Kenya army', *Daily Monitor* (Kampala), 10 September 2011.

58 Author's interview with Ethiopian government minister, Addis Ababa, April 2015.

59 Author's interview with senior Ugandan and AU military and diplomatic official, Kampala, July 2017.

instead been dismissed or vilified. This has manifested itself particularly in strong, collective regional rejection of the notion that politics based around the unit of the clan represents a legitimate or recognisable building block of statehood. One senior Ethiopian foreign ministry official interviewed in May 2013 bemoaned the difficulties of establishing 'government' in Somalia as a result of 'internal problems – the politics of the clan', while a senior Ugandan military official at the AU made a similar observation later that day: 'to reinstitute government [in Somalia] we have to be patient … the people still know their clan as their government, the clan system stayed there and that is the problem'.[60] Museveni argued to counterparts at a 2017 London conference on Somalia that one of the key impediments to state-building in Somalia has been 'the bankrupt ideology of clanism' and that there was a need to 'form political parties with a patriotic, national outlook'.[61] This is not to say that AMISOM forces have not needed to engage in the mediation of clan disputes in their everyday operations – as Djiboutian troops did in Deefow in January 2014[62] – but rather that clan politics has been presented and approached as an obstacle to state-building by AMISOM elites, rather than as part of the structure of political authority in Somalia.

This rejection of 'clanism' derives, in part, from regional elites' connecting of clans to Somali irredentism and their association with Islamist politics, in the context of the GWOT.[63] It also speaks to more regime-specific concerns regarding the dangers of sectarianism and ethnic mobilisation to peace, security and prosperity. This was a key issue that both the current Ethiopian and Ugandan regimes fought against as rebel movements and which they sought to fend off in new post-liberation constitutions, both adopted in 1995.[64] The clan nevertheless represents, as I.M. Lewis notes, perhaps the core political unit in Somali communities and has, in the case of Somaliland, been partly incorporated into the political system as a result.[65]

---

60 Author's interview with senior Ethiopian Ministry of Foreign Affairs official, Addis Ababa, May 2013; Author's interview with senior Ugandan military official, Addis Ababa, May 2013.
61 State House Uganda, 'President Museveni's statement'.
62 AMISOM, 'AMISOM chief of staff presiding clan mediation', January 2014, available at http://amisom-au.org/2014/01/amisom-chief-of-staff-presiding-clan-mediation/.
63 Melito, 'Building terror'.
64 Lovise Aalen, *The politics of ethnicity in Ethiopia: Actors, power and mobilisation under ethnic federalism* (Leiden: Koniklijke Brill, 2011); Nelson Kasfir, 'No-party democracy in Uganda', *Journal of Democracy* 9:2 (1998), 49–63.
65 I.M. Lewis, *A pastoral democracy: A study of pastoralism and politics among the northern Somali of the Horn of Africa* (London: International African Institute, 1961), 161–77; Marleen Renders, 'Appropriate "governance-technology"? – Somali clan elders and institutions in the making of the "Republic of Somaliland"', *Afrika Spectrum* 42:3 (2007), 439–59.

Seeking to develop a Somali state that ignores the reality of clans' profound significance is deeply problematic. The same is true in assuming that relationships between identity and politics in one's own polity will play out in the same way in another – as regional actors appear to have done with the unit of the clan in their thinking on Somalia.

Furthermore, the only version of political authority that has credibly claimed control of south/central Somalia since 1991 – the ICU – was portrayed as unacceptable by regional actors owing, in part, to its lack of conformity to regional governments' views on the place of religion in politics. In explaining their country's decision to rout the Courts through military intervention, Ethiopian Prime Minister Meles told a US journalist in late 2006 that the Courts were 'not interested in democratic, secular government in Somalia', while Foreign Minister Seyoum told Somali officials and Ugandan forces in Mogadishu in mid-2007 that 'we need to be certain that such Islamic elements will not disturb either the Somali government or the Somalia population'.[66] One of Uganda's most senior military officials noted a few years later that 'the Courts had to go … there is no place for religious rule in Africa today'.[67]

Once again, there is a question here as to whether this is regional actors defending local, 'African' norms or whether it is the imposition of one idea of how politics and religion should interact upon a neighbouring polity. Historically, Islamic populations have been largely excluded from politics in Ethiopia, Kenya and Uganda. The three states' rulers have rarely shrunk from expounding their Christian identities or presenting Muslim communities as sources of domestic instability or extremism.[68] Islam is nevertheless embedded in the everyday lives and politics of those in the Somali territories and seeking to promote a 'secular' state in this context fundamentally misunderstands – or wilfully ignores – the place of religious belief and tradition in indigenous Somali structures of political authority.

---

66 Aweys Osman Yusuf, 'Ethiopian foreign minister says troops will stay', *Shabelle Media Network* (Mogadishu), 29 May 2007; 'Transcript: Interview with Meles Zenawi', *Washington Post*, 14 December 2006 .
67 Author's interview with senior Ugandan People's Defence Force official, Kampala, April 2013. These characterisations of the ICU have also, of course, been calibrated in line with GWOT discourses. See Fisher and Anderson, 'Authoritarianism and the securitisation of development'.
68 Jeff Haynes, 'Islam and democracy in East Africa', *Democratization* 13:3 (2006), 490–507.

Conclusion: Reinventing a Regional 'Problem'

This chapter has argued for the importance of analysing state failure, intervention and international relations in Africa through the lens of the region. Both policy-focused and more critical scholars of these phenomena have tended to distinguish 'international' policies, discourses, practices and models from those of the 'local' or 'African'. The case of AMISOM, however, demonstrates the importance of blurring and complicating this binary and interrogating not only how UN and international actors construct, rationalise and engage in military and peacekeeping/peacebuilding interventions, but also how actors at the regional level do so.

This study has demonstrated that AMISOM represents a form of 'African solution' to an 'African problem'. This is not, however, a solution derived from regional actors' empathetic consideration for, or understanding of, existing legitimate forms of Somali political authority – as regional leaders purport in discourses on 'African solutions to African problems'. Instead, it is one which rests firmly on the idea that the Somali crisis can be ended through the imposition of a particular kind of state from the outside. The vision of this state is informed by regional elites' internalisations of Eurocentric, neo-Weberian institutionalist theory on the one hand and transference of their own domestic governance preferences and experiences onto the Somali context on the other.

The chapter has further established the relationship between the longstanding regional image of Somalia as threatening, unpredictable and structurally opposed to the postcolonial contours of African statehood and contemporary regional elites' use of language on failed states, lawlessness and responsibility in their rationalisation of the AMISOM mission. AMISOM, and the language that has cleared a path for it, ultimately constitutes a 'disciplinary' form of regional governance of a kind that has a long history.

These findings challenge not only the legitimacy claims advanced by African elites regarding regional intervention missions, but also those offered by the governments and organisations that finance them. It has become commonplace for US, UK and UN officials in particular to justify increased support for regional peacekeeping operations in terms of supporting locally devised 'African solutions'. Senior UN officials have also emphasised the importance of investing more heavily in regional peacekeeping forces, owing to the 'political leverage' regional states can bring to conflict resolution efforts.[69] This chapter problematises the assumption that geographical proximity equates to an appreciation or respect for local forms of political authority in peacekeeping operations.

Regions have their own political economies and contain multiple contested ideas of what states should look like and how they should relate to one another. It is perhaps unsurprising that political elites should view a regional security crisis through the lens of their own domestic experiences and historical relationships with that state. The AMISOM case raises important questions regarding how far other African-led peacekeeping operations are informed by similar assumptions on the part of regional intervenors. For, while AMISOM and Somalia are perhaps extreme cases, they are not unique. At the time of writing, neighbours are militarily committed to peacekeeping operations of various colours across the continent – from eastern Congo to Nigeria and from South Sudan to Mali. While many of these interventions are premised upon containing or eliminating regional security threats, the AMISOM example underlines the necessity of analysing more closely the broader renegotiations of regional power politics that such interventions enable.

References

Fisher, J. 'Managing donor perceptions: Contextualizing Uganda's 2007 intervention in Somalia', *African Affairs* 111:444, 2012.

Fisher, J. "When it pays to be a fragile state': Uganda's use and abuse of a dubious concept', *Third World Quarterly* 35:2, 2014.

Fisher, J. and D.M. Anderson. 'Authoritarianism and the securitisation of development in Africa', *International Affairs* 91:1, 2015.

Grimm, S., N. Lemay-Hébert and O. Nay. '"Fragile states": Introducing a political concept', *Third World Quarterly* 35:2, 2014.

Hagmann, T. and M. Hoehne. 'Failures of the state failure debate: Evidence from the Somali territories', *Journal of International Development* 21:1, 2009.

Haynes, J. 'Islam and democracy in East Africa', *Democratization* 13:3, 2006.

Herbst, J. *States and Power in Africa: Comparative Lessons in Authority and Control.* Princeton NJ: Princeton University Press, 2000.

Hills, A. 'Somalia works: Police development as state-building', *African Affairs* 113:450, 2014.

---

69 'Evolving peacekeeping landscape requires stronger global-regional partnership, Security Council told', *UN News*, 18 August 2015, available at https://news.un.org/en/story/2015/08/506722-evolving-peacekeeping-landscape-requires-stronger-global-regional-partnership/.

Kaplan, E. *Somalia's terrorist infestation*. Washington DC: Council on Foreign Relations, 2006.

Lotze, W. and P.D. Williams. *The Surge to Stabilize: Lessons for the UN from the AU's experience in Somalia*. New York: International Peace Institute, 2016.

Mazrui, A. 'The blood of experience: The failed state and political collapse in Africa', *World Policy Journal* 12, 1995.

Menkhaus, K. 'State failure and ungoverned space', in *Ending Wars, Consolidating Peace: Economic Perspectives*. London: International Institute for Strategic Studies, 2008.

Menkhaus, K. 'State failure, state-building, and prospects for a 'functional failed state' in Somalia', *The Annals of the American Academy of Political and Social Science* 656:1, 2014.

Migdal, J. *Strong Societies and Weak States: State-society Relations and State Capabilities in the Third World*. Princeton NJ: Princeton University Press, 1988.

Mulugeta, K. *The role of regional and international organizations in resolving the Somali conflict: The case of IGAD*. Addis Ababa: Friedrich Ebert Stiftung, 2009.

Ncube, M. and B. Jones. *Drivers and dynamics of fragility in Africa*. Abidjan: African Development Bank, 2013.

Paris, R. *At War's End: Building Peace after Civil Conflict*. Cambridge: Cambridge University Press, 2004.

Rotberg, R. 'Failed states in a world of terror', *Foreign Affairs* 81:4, 2002.

Rotberg, R. *When States Fail: Causes and Consequences*. Princeton NJ: Princeton University Press, 2003.

Samatar, A.I. 'Destruction of state and society in Somalia: Beyond the tribal convention', *Journal of Modern African Studies* 30:4, 1992.

Soares de Oliveira, R. and H. Verhoeven. 'Taming intervention: Sovereignty, statehood and political order in Africa', *Survival* 60:2, 2018.

Williams, P.D. 'After Westgate: Opportunities and challenges in the war against Al-Shabaab', *International Affairs* 90:4, 2014.

Williams, P.D. and A. Boutellis. 'Partnership peacekeeping: Challenges and opportunities in the United Nations–African Union relationship', *African Affairs* 113:451, 2014.

5

THE RISE OF STRONG MILITARIES IN AFRICA

Defying the Odds?

Jahara Matisek

Introduction

Africa is full of weak states. This problem is generally a product of leadership conspiring to weaken them further for personal gain. This was reaffirmed by the 2016 Fragile States Index (FSI), which noted that only one African country – Mauritius – is rated as stable or higher. Additionally, 26 African states fall into the categories of 'Warning', 'Elevated Warning' and 'High Warning', while the remaining 27 African countries rate as 'Alert', 'High Alert' or 'Very High Alert' (the worst category).[1] Even from a contextual African perspective, the 2016 Ibrahim Index of African Governance (IIAG) noted that the quality of overall governance on the continent has barely increased one point since 2006, while most states have suffered increases in corruption, alongside a deterioration in accountability, safety and the rule of law.[2]

The issue of state weakness had primarily gone unnoticed by the international community prior to the 9/11 terrorist attacks. However,

---

1   See the Fragile States Index, 2016, available at https://fundforpeace.org/2016/06/27/fragile-states-index-2016-the-book/.
2   'The 2016 Ibrahim Index of African Governance: Key Findings', available at http://mo.ibrahim.foundation/news/2016/progress-african-governance-last-decade-held-back-deterioration-safety-rule-law.

the attacks prompted the US and the international community to address weak states because they were viewed as harbouring terrorists. The George W. Bush administration identified state weakness as a high-level priority in 2003, with it being a focal point of President Bush's first National Security Strategy (NSS).[3] However, the international focus on solving the weak state problem has made little progress, despite high levels of engagement and investment, because much of this has been unable to overcome local-level politics and other informal institutions and factors that prevent the weak state from strengthening. Many international interventions have failed to address the structural problems and grievances that make insurgency and terrorism more likely across Africa, the Middle East and Asia. States such as Liberia and Sierra Leone appear to be escaping the weak state problem under international trusteeship and occupation; if peacekeepers and foreign donors were to leave, both countries would likely collapse.[4]

A recent article in the journal *The American Interest* notes that, not only do weak and failing states provide permissive environments for terrorism, but they also create cascading levels of risk to neighbouring states, thus destabilising the international order.[5] This is partly driven by the issue of weak state political contexts where elected leaders are merely a façade. The real power brokers (for example, local 'Big Men') operate informally at the local level with militias that are more powerful than the state army, allowing them to exert authority, legitimacy and power. The collapse of Libya after the UN-sanctioned military intervention in 2011 and the continued disorder and chaos there, described by Islam Goher in Chapter 7 of this volume, illustrates the crux of this issue and its latent effects. At the same time, external actors attempting to solve and alleviate state weakness (and the challenges it poses) have found it difficult; this may even create more damaging externalities.[6] For instance, research into Western security force assistance (SFA) provided in weak state political contexts indicates the haphazard and *ad hoc* nature of trying to improve

---

3   See Susan E. Rice, 'The new national security strategy: Focus on failed states', Brookings Institution *Policy Brief* 116, 2003, available at https://www.brookings.edu/wp-content/uploads/2016/06/pb116.pdf.

4   For discussion, see Anders Themnér (ed), *Warlord Democrats in Africa: Ex-Military Leaders and Electoral Politics* (London: Zed Books, 2017).

5   Seth Kaplan, 'Weak states: When should we worry?', *The American Interest* 12:4 (2017), available at https://www.the-american-interest.com/2017/01/26/weak-states-when-should-we-worry/.

6   Stuart E. Eizenstat, John Edward Porter, and Jeremy M. Weinstein, 'Rebuilding weak states', *Foreign Affairs* 84:1 (2005).

host nation security institutions when the recipient government cannot afford this new army, views it as a threat or uses it to eliminate rivals to consolidate power.[7]

The traditional response by the international community to endemic and systemic problems in weak states has been to increase economic aid, send humanitarian non-governmental organisations (NGOs) and/or attempt to strengthen the security forces through SFA.[8] However, these external resources have been a double-edged sword. Famously, Dambisa Moyo argued in *Dead Aid* that decades of 'systemic aid' created a culture of aid dependency on the African continent, fostering the development of corrupt politics, which entrenched patrimonial regimes.[9] Moyo's argument follows a similar logic established over decades of research on the ills of foreign aid, from William R. Easterly to Péter Tamás Bauer, where Western aid to developing states does more harm than good.[10] This aid creates moral hazard by subsidising leaders that make poor choices, do just enough to keep the international aid flowing, but not enough to address the root causes of their weak state. Even when international aid becomes conditional on domestic reforms, this can be difficult for a leader to implement or sustain in the long term because economic growth and tax revenue collection may lag, leading to underpayment to informal power brokers who may then rebel against the government as patronage is reduced.

Worse yet, SFA has sometimes enabled the host military to take control of the government through a *coup d'état*, or has given the armed forces the capacity and rationale to dominate domestic politics and repress dissidents through the constructed veneer of labelling opponents as 'terrorists'.[11] In other cases, such as Iraq, Afghanistan and Somalia, Western military assistance created 'Fabergé Egg' armies – expensive host militaries that are easily 'cracked' by insurgents.[12] Making matters worse, throughout the

---

7 Jahara Matisek and William Reno, 'Getting American Security Force Assistance Right: Political Context Matters', *Joint Force Quarterly* 92:1 (2019), 65–73; William Reno, 'The politics of security assistance in the horn of Africa', *Defence Studies* 18:4 (2018), 498–513.

8 Ashraf Ghani and Clare Lockhart, *Fixing Failed States: A Framework for Rebuilding a Fractured World* (New York: Oxford University Press, 2009).

9 Dambisa Moyo, *Dead Aid: Why Aid is Not Working and How There is a Better Way for Africa* (New York: Macmillan, 2009).

10 William Russell Easterly, *The White Man's Burden: Why the West's Efforts to Aid the Rest Have Done So Much Ill and So Little Good* (New York: Penguin, 2006); Péter Tamás Bauer, *Dissent on Development* (Cambridge MA: Harvard University Press, 1976).

11 Abraham Kaplan, *The Counter-terrorism Puzzle: A Guide for Decision Makers* (New York: Routledge, 2017).

12 Jahara Matisek, 'The crisis of American military assistance: strategic dithering and Fabergé Egg armies', *Defense & Security Analysis* 34:3 (2018), 267–90.

African Sahel, states appear mired in perpetual internal conflicts – driven by climate change and smuggling networks – further weakening the state and its security apparatus.[13]

During the Cold War, many regimes in weak states could rely on external military aid and assistance to repress citizens and/or fight domestic insurgencies and still be successful in retaining power, such as in Angola. Other states, such as Egypt, have allowed the government and society to be politically devoured by the military. Egypt has been the archetypal model of military dominance in the political sphere since 1952, when a group of Egyptian military officers removed King Farouk. From that point, the political system and other parts of the bureaucracy and government have been dominated by military elites that were primarily educated and trained in the West.[14] Not every state has allowed its military to dominate. For instance, some African states have created institutionally strong militaries that have never threatened the political leadership, such as in Botswana, Namibia and Djibouti. These are anomalies in Africa, however, where such positive civil-military relations are not the norm.

Phillip Roessler notes that fearful political leaders in sub-Saharan Africa (SSA) typically manage their militaries by coup-proofing themselves through public purges, ethnic exclusion or by creating multiple parallel security institutions and presidential guards. These efforts reduce coup attempts by increasing loyalty, but also undermine military effectiveness. Such a systematic strategy is not perfect, since historical evidence shows that purging a former ally increases the chance of civil war.[15] However, purges of high-ranking officers or those that undertake anti-regime activities has been shown to reduce the chances of civil war, especially when it is perceived publicly as justified, such as the 1984 purges in Cameroon or the dismissal of top-ranked Angolan military officers in 2002.[16] Any attempt to purge or diminish various actors in enumerable security institutions presents risks, which a leader must constantly weigh against the doling out of patronage.

---

13 See Buddhika Jayamaha, Jahara Matisek, William Reno, and Molly Jahn, 'Climate Change and Civil War Dynamics: Institutions and Conflicts in the Sahel', *Journal of Diplomacy* forthcoming 2019.

14 The 2013 Egyptian military coup was led by Army General Abdel Fattah el-Sisi, who had attended the US Army War College, in Carlisle, Pennsylvania, in 2006. General Sisi removed Egypt's first freely elected President, Mohamed Morsi, on the grounds that he was allegedly trying to create a religio-fascist government.

15 Phillip Roessler, 'The enemy within: Personal rule, coups, and civil war in Africa', *World Politics* 63:2 (2011), 300–46.

16 Jessica Maves Braithwaite and Jun Koga Sudduth, 'Military purges and the recurrence of civil conflict', *Research and Politics* 3:1 (2016), 1–6.

In this vein, the elimination of allies, inclusion of rivals and shuffling of high-ranking officers is a balancing act where, if the ruler allows one part of the military to become effective, it will potentially increase the chances of their own overthrow.

Creating a strong military goes against the traditional scholarly wisdom espoused by Jackson and Rosberg. They contend that the international system is responsible for keeping African states 'alive on paper' (for example, in terms of juridical sovereignty) despite them lacking the ability to meet the basic empirical criteria for statehood (such as providing basic goods and services for their citizens). Jackson and Rosberg further conclude that, whenever there is some modicum of military strength and capacity, it either reflects politicisation of military institutions or that the military is running the state.[17] However, how can such weak states create strong militaries despite lacking resources, strong governmental institutions and/or traditional state capacity? What formal and informal relationships exist between political leadership and their militaries that permit coexistence between one and the other? Are internal and external threats perceived sufficiently to drive a regime leadership to develop a security apparatus stronger than the rest of the state? These questions drive us towards the puzzle of how a low-capacity state might create a high-capacity military, as most of the existing academic literature suggests that a weak and poor state will have a weak military.[18]

If we look beyond Jackson and Rosberg's identification of 'stateness', there may be an alternative process for a poor and underdeveloped state becoming an *exercitu civitatis* ('army state'). Such states appear to have emerged around the end of the Cold War, where the military is strong, loyal to the state and citizenry and is not formally in charge of the government. Specifically, as this chapter will show, Senegal, Uganda, Rwanda and Ethiopia have defied conventional logic in Africa by building capable militaries that are not a threat to the government (or public) because these armies have been integrated into the nation-building programme of the state. Through ideologies and political programmes, each of these states have also integrated the military in such a way that it does not undermine military effectiveness.

---

17 Robert H. Jackson and Carl G. Rosberg, 'Why Africa's weak states persist: The empirical and the juridical in statehood', *World Politics* 35:1 (1982), 1–24.

18 For instance, a seminal work on civil wars by Fearon and Laitin (2003) suggests that a weak state has a weak military and this is premised on a weak economy: James D. Fearon and David D. Laitin, 'Ethnicity, insurgency, and civil war', *American Political Science Review* 97:1 (2003), 75–90.

This gives the army the necessary space to be a bureaucratically competent enclave in a patrimonial state – thereby giving military elites a motivational purpose for partnering in state-building operations.

Reconsidering Strong Militaries

Well before the advent of modern industrialised warfare, ideas and conceptions surrounding what makes a military 'strong' – and its impact on society – have abounded. In Plato's *The Republic*, Socrates contended that warfare improved the welfare of cities; maintaining strong and reliable city guardians was dependent on keeping them physically fit, savage and more educated than their adversaries.[19] Much of Sun Tzu's thinking on military power in *The Art of War* relied on the ability of commanders to properly plan and outsmart opponents in battle. However, this capability was contextually dependent on generals being subservient and loyal to the state.[20] Outside of traditional military literature, Adam Smith even suggested in a lesser-read chapter of *The Wealth of Nations* that, in order for industrialised societies to grow and remain competitive, they needed to increase specialisation – to include military structures and institutions. For such a state to remain economically successful, taxation would be required to create and maintain standing armies specialising in warfare for the defence of the nation and its resources.[21] To Smith, state capacity to collect tax revenues begat military capacity as well.

Napoleon Bonaparte illustrated how the proper organisation of a military, especially promotions based on merit, could create highly capable armies that were more manoeuvrable and destructive.[22] Clausewitz, writing in *On War*, reasoned that political context mattered most and that whichever state could generate more resources and employ them correctly would have a superior military. However, he cautioned that battlefield

---

19 Allan Bloom and Adam Kirsch, *The Republic of Plato* (New York: Basic Books, 2016). For discussion of armies in ancient times see also Richard Carpenter, 'The Military Character of Plato's Republic', (University of Auckland master's thesis, 2010).

20 Sun Tzu, *The Art of War: Sunzi's Military Methods*, trans Victor H. Mair, (New York: Columbia University Press, 2007).

21 Adam Smith, *An Inquiry into the Nature and Causes of the Wealth of Nations* (London: Strahan & Cadell, 1776), Book V, Chapter 1.

22 See Peter J. Dean, 'Napoleon as a military commander: The limitations of genius', *The Napoleon Series*, 2000, available at http://www.napoleon-series.org/research/napoleon/c_genius.html.

success required military leaders to possess 'genius', which would enable them to overcome the fog and friction of war. Clausewitz went on to define the more tangibly important war-making resources as *'the fighting forces proper, the country*, with its physical features and population, and its *allies'* (emphasis in the original), which required political leadership to give authority to military leaders to best achieve military goals within the framework of desired political ends and objectives.[23]

At the beginning of the Cold War, Hans Morgenthau, observing the link between industrialisation and the capacity to wage war, theorised that the power of a military was a product of geography affording natural defences, such as oceans and rough terrain. Land also provided the necessary raw materials for economies to build armies. To Morgenthau, such geographical blessings could be translated into economic power, which profoundly influenced the quantitative and qualitative capabilities of a military, such as with the rise of Great Britain and later the US, David Singer (and his team) were the first to operationalise such theorising into measurable constructs, establishing the Correlates of War (COW) project, which made six variables important for national military power, known as the Composite Index of National Capabilities (CINC): total population, urban population, iron/steel production, energy consumption, military expenditure and number of personnel in the military.[24]

Over time, the COW project has updated annual CINC scores for each state dating back as far as 1812 (or their entry into the international system). However, various scholars have disputed the role of CINC in dictating military strength and war outcomes, especially because war is generally a struggle of political willpower (such as with American military involvement in Vietnam) and chance. Others have found better material measures of military power, known as Dispute Outcome Expectations (DOE), which rely on 'ensemble learning' to better predict military strength/capability in a conflict.[25] Despite the growing list of military indices and models that should help predict battlefield outcomes, these fall short when explaining military effectiveness on the African continent. For example, economic power could not have predicted

23 Carl Von Clausewitz, *On War*, trans,. Michael Howard and Peter Paret, (Princeton NJ: Princeton University Press, 1989), 78–80.

24 David J. Singer, Stuart Bremer, and John Stuckey, 'Capability distribution, uncertainty, and major power war, 1820–1965' in Bruce Russet (ed.), *Peace, War, and Numbers* (Beverly Hills CA: Sage, 1972), 19–48.

25 Robert J. Carroll and Brenton Kenkel, 'Prediction, Proxies, and Power', *American Journal of Political Science* forthcoming 2019.

the dismal performance of the far richer and superior Libyan military – on paper at least – in 1987. The smaller and under-resourced Chadian military armed mainly with Toyota 'technicals' (pickup trucks retrofitted with anti-tank weapons and heavy machine guns) overran the Libyan military on the Chadian border, killing over 7,000 Libyan troops and destroying or capturing almost $1.5 billion in Libyan military equipment.[26] This Libyan military disaster stemmed from coup-proofing strategies, such as preventing the Libyan armed forces from training effectively and also the insertion of loyal – but incapable – commanders into important combat positions.[27]

The issue of overly material predictions of military effectiveness is why a US Government Accountability Office (GAO) report noted numerous 'problems of measuring military capability and the difficulty of quantifying military capability in a single, definitive measure'. This GAO report highlighted the need to assess and evaluate military capacity and strength based on four variables: force structure, modernisation, sustainability and combat readiness.[28] This has not stopped scholars, such as Organski and Kugler, from advocating population size as being important for military capacity and asserting the wealth of the state is responsible for creating military strength. They further contend this determines the level of technologically expensive weapon systems that can be employed by the military.[29] Indeed, Fearon and Laitin's seminal article on civil wars contended that economic productivity – especially higher per capita incomes – are strong indicators of military and state capacity.[30] More recently, Phil Arena created an index of military strength based on modernisation of the military (known as the M-Score), which takes the number of military personnel in a state and divides it by its military budget. This has enhanced the validity of such metrics to predict war outcomes.[31]

The problem with these materialistic assumptions is that their explanatory power is only useful in explaining the military strength (and

26 J. Millard Burr and Robert O. Collins, *Africa's Thirty Years' War: Chad-Libya-the Sudan, 1963-1993* (Boulder, CO: Westview Press, 1999); Kenneth M. Pollack, *Arabs at War: Military Effectiveness, 1948-1991* (Lincoln NE: University of Nebraska Press, 2004).

27 Florence Gaub, 'The Libyan armed forces between coup-proofing and repression', *Journal of Strategic Studies* 36:2 (2013), 221–44.

28 US Government Accountability Office, *Measures of Military Capability: A Discussion of Their Merits, Limitations, and Interrelationships*, NSIAD-85-75 (Washington DC: US Government Accountability Office, 1985), available at http://www.gao.gov/assets/150/143000.pdf.

29 A.F.K. Organski and Jacek Kugler, *The War Ledger* (Chicago IL: University of Chicago Press, 1981).

30 Fearon and Laitin.

31 Phil Arena, 'Measuring Military Capabilities', (University of Essex Working Paper, 2016).

success) of large industrial states that are major players in the international system. However, in the developing world, where economies are unstable due to shifting commodity prices and availability, the capacity of the state is uneven and coup-proofing strategies typically undermine military capacity. Thus Gupta, de Mello and Sharan's work indicates that underdeveloped states with high military spending are positively correlated with higher levels of corruption and that these weak states likely purposively misreport military spending or use high expenditures for patronage.[32]

While economic power may facilitate the creation of a modern military, it does not guarantee victory on the battlefield. As Kenneth Waltz once put it, 'Inability to exercise *political* control over others does not indicate *military* weakness' (emphasis in the original).[33] Assumptions about a technologically advanced military being able to defeat small and poor adversaries were recently disproven by Lyall and Wilson. They demonstrated that when industrialised militaries increased mechanisation over the period 1800–2005, their ability to defeat insurgents actually *decreased* because counterinsurgency forces were less likely to interact with locals face-to-face, reducing intelligence collection.[34] This is a crucial flaw in most Western styles of warfare, as high-fidelity intelligence is a vital resource in trying to win any type of civil war.[35]

Wealth, technological advantages and military budgets are troublesome predictors of military capacity and ability in Africa. Stephen Biddle has contended that qualitative factors within the military matter more. To that point, Biddle writes in *Military Power* that 'force employment, or the doctrine and tactics by which armies use their materiel in the field' is a more important measure of military capacity than other material explanations (for example, COW/CINC, economic GDP, and so on).[36] Digging deeper into non-material explanations of military power, Caitlin Talmadge argues that political context internally and historically – basically, civil-military relations – matter in how authoritarian militaries

32 Sanjeev Gupta, Luiz de Mello, and Raju Sharan, 'Corruption and Military Spending' in George T. Abed and Sanjeev Gupta (eds), *Governance, Corruption, & Economic Performance* (New York: International Monetary Fund, 2002).

33 Kenneth N. Waltz, 'International Structure, National Force, and the Balance of World Power', *Journal of International Affairs* 21:2 (1967), 227.

34 Jason Lyall and Isaiah Wilson, 'Rage against the machines: Explaining outcomes in counterinsurgency wars', *International Organization* 63:1 (2009), 67–106.

35 See Stathis N. Kalyvas, *The Logic of Violence in Civil War* (New York: Cambridge University Press, 2006).

36 Stephen Biddle, *Military Power: Explaining Victory and Defeat in Modern Battle* (Princeton NJ: Princeton University Press, 2010), 2.

fight. Usually, political leaders see value in keeping their state institutions and army weak, because they are more worried about being overthrown than losing a war. Talmadge reasons that when militaries have control over promotion systems, training regimens, command arrangements and information management, they fight better than under regimes that try to control and manipulate these four variables.[37]

Given that most African states are hybrid regimes – a blend of authoritarianism, democracy and patrimonialism – that are resource-poor and underdeveloped, should it be a surprise that some regime leaders decide to undermine their institutions because they might threaten their rule?[38] Perhaps it is better to reconsider what the implications are for a state that allows a strong military to exist as a bureaucratic enclave in a sea of patrimonialism. More importantly, there are internal and external dynamics that can facilitate the creation of certain types of 'strong' militaries in weak states, but, and this is crucial, this is highly dependent on regime leadership strategically tying its military to nation- and state-building. Without the necessary political context, no amount of foreign aid or SFA can artificially create a viable army in a weak state that the recipient political system sees no value in creating or sustaining. Constructive civil-military relations are necessary for an effective military to emerge in weak states, where it is only possible through strategic partnerships between political, societal and military elites.[39] This defies traditional Western civil-military relations logic, which rests upon strict civilian control of the military and an army separated from politics.[40]

The Role of Militaries in the State: The Rise of the *Exercitu Civitatis* in Africa?

The most important basis of a state rests on its Weberian ability to monopolise coercion and violence and maintain territorial sovereignty. For much of

---

37 Caitlin Talmadge, *The Dictator's Army: Battlefield Effectiveness in Authoritarian Regimes* (Ithaca NY: Cornell University Press, 2015).

38 See Joel S. Migdal, *Strong Societies and Weak States: State-Society Relations and State Capabilities in the Third World* (Princeton NJ: Princeton University Press, 1988).

39 Rocky Williams, 'Towards the Creation of an African Civil-Military Relations Tradition', *African Journal of Political Science / Revue Africaine de Science Politique* 3:1 (1998), 20–41.

40 See for instance Huntington (1957), which is revered as the seminal work on Western civil-military relations in terms of how militaries should be controlled by civilian leadership and how they should be subservient: Samuel P. Huntington, *The Soldier and the State: The Theory and Politics of Civil-Military Relations* (Cambridge MA: Harvard University Press, 1957).

history, the ability of a polity to exist was fundamentally tied to the ability of its leaders to mobilise an army for offensive and defensive purposes. This supports Charles Tilly's maxim that 'War made the state, and the state made war'.[41] Tilly's statement logically tied together the obvious need for a military – to include some modicum of state capacity – to be effective in waging offensive wars and in defending sovereign territory. To maximise successful war efforts, the polity needed to efficiently develop institutions and bureaucracies capable of supporting the military in terms of logistics, which also required the state to devise ways to tax and collect revenues to support military operations. However, the strength and power of the military did not guarantee survival of the capital city, as such a warrior class created a novel political problem due to their presence. The threat posed to the political system by the existence of a military is what Peter Feaver has called the 'civil-military problematique'.[42]

Socrates was the first to astutely recognise, in Plato's *The Republic*, the need for an army to defend the city from foreign invaders, but also that such 'guardians' and savagery posed a threat to the people in the city. The only solution to this political problem was to make the guardians 'philosophic', so they would not exploit their favourable position.[43] The problem of militaries running state machinery directly (or indirectly) has led to them being identified as sometimes forming a 'Praetorian State'. In 1939, Max Lerner was the first scholar to note the rise of such governments after World War One – primarily in the context of advanced weaponry supporting emerging totalitarian socialist governments and fascist regimes.[44] Similarly, Harold Lasswell warned of the 'Garrison State' in 1941, where military professionals who had become specialists in violence due to the rapid technological advances in warfare – air combat especially – would come to dominate politics and state managerial duties. Laswell reasoned that the skills learned through such technologically complicated war-making would lead military professionals to take over the state.[45] These suggestions about states becoming militarised are worrisome to this day, especially in modernised states. However, most African states do not

---

41 Charles Tilly (ed.), *The Formation of National States in Western Europe* (Princeton NJ: Princeton University Press, 1975), 42.
42 Peter D. Feaver, 'The civil-military problematique: Huntington, Janowitz, and the question of civilian control', *Armed Forces & Society* 23:2 (1996), 149–78.
43 Bloom and Kirsch, 93–146 and 349–50.
44 Max Lerner, *It is Later Than You Think: The Need for a Militant Democracy* (Piscataway NJ: Transaction Publishers, 1939), 44 and 50.
45 Harold D. Lasswell, 'The garrison state', *American Journal of Sociology* 46:4 (1941), 455–68.

have a technologically modern military because of the costs and thus it is less likely that they will become a Praetorian State or Garrison State.

In 1906, Otto Hintze had argued that different types of military build-ups had an impact on classes and domestic politics. Hintze noted that 'sea power is allied with progressive forces, whereas land forces are tied to conservative tendencies'.[46] While he wrote this before the advent of air forces, his logic could extend to the fact that an air force would be even more progressive than sea power, on the assumption that technological advances require improvements in economic and educational conditions. Thus, when considering how richer states tend to invest in technology (such as air and sea capabilities) to permit the creation of smaller armies, resulting in more liberalising forces, then efforts by poor states to build strong militaries with minimal naval assets and aircraft will likely experience less liberalisation. Being unable to afford expensive weapon systems in the sea and air domains likely also reflects an inability to develop the human capital necessary to operate and maintain these complicated weapon systems.[47] Thus, most of Africa is suffering from a stunted form of state-building because military institutions have not modernised, due to a lack of resources and institutions to support modern technological forms and ways of warfighting. This lack of progression means that militaries in Africa tend to be army-centric and thus act as a conservative force in society.

The theoretical expression of *exercitu civitatis* is an important reconceptualisation of previously used terms that described regimes with military elites in charge. Regardless, Plato, Sun Tzu, Clausewitz and Huntington have each shared the belief that it is unnatural for a military to rule society. However, now that building and maintaining technologically advanced militaries is such a cost-prohibitive endeavour, small, poor and underdeveloped states lack the necessary financial resources, human capital and infrastructure to commit to (and maintain) complicated weapon systems for their militaries. This stunts military behaviour in such states. Building a technologically advanced military is a capital and labour-intensive venture. Lacking such internal resources leads to reliance on external assistance, but this is not a long-term panacea. Maintaining

---

46 Otto Hintze, 'Military Organization and the Organization of the State' (1906), 178–215. Reprinted in John Hall (ed.), *The State: Critical Concepts* (New York: Routledge, 1994), 202.

47 While there have not been any statistical analyses to validate this claim, author interviews conducted in 2015–18 with Western military officials that conduct military assistance in weak African states corroborate that these countries lack the human capital, logistics and institutions necessary to support military aircraft and naval vessels.

a modern military requires a significant commitment of resources in the long term, as many advanced weapon systems, such as fourth-generation fighter aircraft, require a substantial commitment of national resources to develop aviators, well-trained maintainers and substantial infrastructure requirements (such as airports that meet the minimum standard for aircraft and logistic support).

The International Institute for Strategic Studies publication, *The Military Balance*, annually lists military specifics and details for all states. It notes that many African states lack either the money, parts, skills or expertise to keep advanced weapon systems operable (this can also sometimes be attributable to sanctions and embargoes).[48] At the same time, air power can be an effective tool in dealing with insurgents primarily because, as a rule of thumb, guerrillas/insurgents/terrorists lack an air force or other aviation-based capabilities.[49] Nonetheless, some countries, such as Uganda and Rwanda, have downsized the numbers and scope of their military aircraft, shifting towards cheap and reliable Russian helicopters as the primary means of projecting air power. Indeed, when the Rwandan military was considering the acquisition of American C-130s in 2016, they experienced 'sticker shock' in discovering that one C-130 would consume a quarter of the Rwandan defence budget, before deciding to pursue cheaper Cessna aircraft.[50] In this case, budgetary limitations have driven a state to seek out optimal air power that fits its political and economic context, while maximising overall military effectiveness.

---

48 See, for example, *The Military Balance 2017* (London: International Institute for Strategic Studies, 2017), 479–548.

49 An exception to this rule is the way in which Islamic State has employed booby-trapped drones in ambushes and relied upon improvised 'bomber' drones that carry/deliver a small mortar round. See Kelsey D. Atherton, 'IED Drone Kills Kurdish Soldiers, French Commandos', *Popular Science*, 11 October 2016, available at http://www.popsci.com/booby-trapped-isis-drone-kills-kurdish-soldiers-french-commandos; Tom O'Connor, 'ISIS has no Air Force, but it has an army of drones that drop explosives', *Newsweek*, 17 April 2017, available at http://www.newsweek.com/isis-air-force-army-drones-drop-bombs-585331; Besides kinetic attacks, drones can serve as a powerful propaganda tool, as with the Taliban using cheap drones to film their successful attacks against the Afghan government and security forces. See Austin Bodetti and Hamidullah Barakzai, 'Here Come the Taliban Drones', *The Diplomat*, 2 November 2016, available at http://thediplomat.com/2016/11/here-come-the-taliban-drones/. Finally, the Tamil Tigers in Sri Lanka were the first rebel group to organically develop their own air force called the 'Air Tigers' that was a nuisance to the Sinhalese military. See Paige Ziegler, 'Learning from the Liberation Tigers of Tamil Eelam', *The Strategy Bridge*, 13 April 2017, available at https://thestrategybridge.org/the-bridge/2017/4/13/learning-from-the-liberation-tigers-of-tamil-eelam.

50 This point draws on the author's fieldwork and interviews at US Africa Command, Stuttgart, Germany and in Kigali, Rwanda, in August 2017.

As Jeremy Weinstein has argued, a resource-poor military that 'punches above its weight' can be attributed to its army having developed its qualitative war fighting abilities.[51] Thus, it seems that a weak state with a strong military ends up favouring a large army with military equipment that is cheap to use and maintain. When it comes to more complicated weapon systems, such states tend to rely on contractors to fill the technical 'gap' (for example, lack of capable indigenous personnel), as seen in places such as Sierra Leone, Liberia and Nigeria.[52]

In other cases, the strategic use of private military companies (PMCs) and mercenaries is due to a regime not wanting their military to be too capable, such as in Saudi Arabia, Oman and Libya under Colonel Muammar Gaddafi. More importantly, since the typical African state is not primarily concerned with the prospect of conventional inter-state warfare, it chooses to develop armies that are cheap and combat-effective in dealing with rebels and other regional threats, to include the spill-over effects of civil wars in neighbouring states, such as refugees. Such 'spill-over' can even be beneficial, as a weak state with a capable army can seek out revenues from the international community, such as the UN or African Union (AU), by providing peacekeeping services.

Perhaps the reason the 'weak states with strong militaries' narrative is not discussed more in contemporary debates and the current literature is because it was a common state-building formula pursued during the Cold War. However, the *exercitu civitatis* phenomena did *not* exist at that time, because most African militaries were so heavily subsidised by foreign patrons providing advanced and expensive weapon systems. Thus, such states with complicit approval from their patron state had the political willpower (and external support) to employ repressive domestic policies that kept a lid on civic society through the suppression of its citizenry, all in the name of defeating communism (or 'western democracy'). However, as the Soviet Union began to fracture in 1989, many client states that had survived based on that security clientelism relationship, began to fragment as well, such as the Derg regime in Ethiopia. Indeed, even some western-backed 'client' states collapsed, such as in the case of Somalia (despite it having one of the largest militaries in Africa at the time), because the US no longer had a rationale to support anti-Soviet client states in Africa

---

51 Jeremy M. Weinstein, *Inside Rebellion: The Politics of Insurgent Violence* (New York: Cambridge University Press, 2006).
52 Karin Dokken, *African Security Politics Redefined* (New York: Palgrave Macmillan, 2008).

and elsewhere. Indeed, new forms of civil war erupted (or intensified) as the Cold War ended, where states – that had been held together through coercion – became untenable without a strong patron state, leading some regions like West Africa and the Balkans to fracture.[53] While it might seem easy to blame the US and Soviet Union for the emergence of civil wars after the Cold War therefore, leaders in many of these states made agential decisions to create political systems and institutions that fuelled ethnic tensions. External aid merely subsidised these bad political decisions.[54]

Resourceful and Ideational: The Strong Militaries of Senegal, Ethiopia, Uganda and Rwanda

When it comes to military strength and capacity in Africa, this should be considered in terms of relativity. There are various structural constraints facing the development of a military in African states, especially when there is an attempt to adapt structures and institutions from Western or Eastern doctrines.[55] Western doctrines emphasise expensive weapons, command and control systems and decentralised authority at the tactical level. Eastern doctrines are generally best characterised by the aphorism about the Soviet way of war attributed to Josef Stalin: that quantity has a quality all its own. Eastern doctrines are much more focused on waging a war in a centralised fashion. Unfortunately, neither approach fits African militaries neatly, where most states lack the resources for a Western-style military, while lacking the population density to support a large standing military.[56] What if there was an alternative pathway to a strong military that deviates from Western and Eastern conceptions?

Senegal, Ethiopia, Uganda and Rwanda meet the criteria of weak state capacity, but they have also established capable and loyal armies. These four states share the common bond of having a GDP per capita of

---

53 See Mary Kaldor, *New and Old Wars: Organised Violence in a Global Era* (Malden MA: Polity Press, 2012).

54 Donald L. Horowitz, *Ethnic Groups in Conflict* (Berkeley CA: University of California Press, 2000).

55 Herbert M. Howe, *Ambiguous Order: Military Forces in African States* (Boulder CO: Lynne Rienner, 2001), 3.

56 Collier (2008) considers the continent too poor, and Herbst (2014) laments the low population densities in Africa. See Paul Collier, *The Bottom Billion: Why the Poorest Countries are Failing and What Can be Done About It* (New York: Oxford University Press, 2008); Jeffrey Herbst, *States and Power in Africa: Comparative Lessons in Authority and Control* (Princeton NJ: Princeton University Press, 2014).

$600–$1,000 as of 2017, while the average SSA country has a GDP per capita of $1,500.[57] More importantly, they each exhibit aspects of *exercitu civitatis*, where the military is a part of state formation and development and the overall political programme.

Senegal has never experienced a military coup attempt since gaining independence from the French in 1960, and it has also experienced minimal internal strife. Although the Casamance conflict in Senegal ebbed and flowed from 'cold' to 'hot' in 1982–2014, the Senegalese military managed to keep the conflict isolated to the Casamance region, which allowed Senegal to remain stable while political deals were brokered to win over various rebel factions and facilitate their entry into the political system.[58] The strength of the Senegalese military might also lie in its initial years, when the first president, Léopold Sédar Senghor, created harmonious civil-military relations in 1962 through delicate meetings with General Jean Alfred Diallo, agreeing that the military should be used for state-building (that is, modernisation).[59] From that point, the Senegalese military took on a developmental role: from training military doctors to dealing with yellow fever and cholera outbreaks, to creating an engineering corps responsible for building infrastructure.[60] These efforts were accomplished under the auspices of *Armée-Nation*, where the Senegalese armed forces were dedicated towards state development with an emphasis on building close and constructive relations with the public.

Due to the central role of the Senegalese military in society, it has not become a backwater employing the lowest segments of society. Instead, Senegal has attracted the most educated classes from the youth for military service.[61] Finally, according to a military officer at US Africa Command, a unique strength of the Senegalese military has been its resourcefulness. By this he meant that 'they leverage their deployments in support of UN peacekeeping missions to build a more effective military ... make good use of their partners ... [and rely on] a network of training centers throughout the region [which] are used to provide the resources normally sourced nationally for larger and richer countries'.[62]

---

57 See World Bank, 'GDP per Capita', available at http://www.worldbank.org/.

58 These points are based on the author's fieldwork and interviews in Dakar, Senegal, in August 2017.

59 Iba Der Thiam and Mbaye Gueye, *Atlas du Sénégal* (Paris: Éditions Jeune Afrique, 2000).

60 Biram Diop, 'Civil-military relations in Senegal', in Dennis C. Blair (ed.), *Military Engagement: Influencing Armed Forces Worldwide to Support Democratic Transitions* (Washington DC: Brookings, 2013), 236–56.

61 Ibid.

62 US military officer, personal communication with the author, 13 April 2017.

In this sense, the Senegalese military is kept occupied by developing technical and professional qualities, while focusing on a host of issues that do not include domestic politics. This lack of interference by the military, and the decision by the Senegalese president to put his military to work domestically, set it on a path-dependent process for professionalism and focus on technocratic skills. Moreover, the decision by then-French President Charles de Gaulle to engage in a form of neo-colonial institutional relations helped retain close political and military links with Senegal.[63] However, the defining point of French policy was its long-term horizons, which made it easier for Senegalese politicians and military leaders to make short-term sacrifices in exchange for long-term pay-offs.[64] Such relations have provided a strong linkage and reinforced the path dependence of constructive civil-military relations, where the Senegalese military is professional enough (due to its foreign patron link) and retains a harmonious relationship with political leadership and society.

Outside the Western-based model that worked in Senegal, Ethiopia is the most ancient nation-state in the world, though this was briefly interrupted by Italian rule (1936–41). Following occupation, Ethiopia returned to rule by an emperor until the Derg regime, led by communist military officers, staged a coup in 1974. This ended three millennia of monarchical rule over the Kingdom of Abyssinia, and the Derg regime radically changed the nature and character of the Ethiopian state and its institutions. The Ethiopian People's Revolutionary Democratic Front (EPRDF) defeat of the Derg in 1991 in turn led to the creation of the leftist revolutionary democracy ideology of the contemporary Ethiopian government and military. To facilitate the transition from communist military rule, about 5,000 Derg personnel were permitted to keep serving in the new Ethiopian National Defense Force (ENDF), with some limits on command positions in the new government and military, because they had been indoctrinated by the Derg regime. With the ENDF established as a new national army, it fully integrated all ethnic groups, avoiding the pitfalls of an ethnically partitioned multi-national force that would be more likely to engage in a coup.[65] While there was some initial favouritism shown towards ethnic Tigrayans, as of 2018 the ENDF has become more balanced in the enlisted and officer corps after several

---

63 See Georges Lavroff (ed.), *La Politique Africaine du General de Gaulle* (Paris: Pédone, 1990).
64 Shaun Gregory, 'The French Military in Africa: Past and Present', *African Affairs* 99:396 (2000), 435–48.
65 See Kees Koonings and Dirk Kruijt, *Political Armies: The Military and Nation Building in the Age of Democracy* (London: Zed Books, 2002).

iterations of personnel realignments.[66] The shift towards a more balanced military occurred due to the British military assisting the ENDF with reforms in 2004–5, which sought to learn from the disastrous border war with Eritrea (1998–2000).[67] Finally, the 'guerrilla mindset' that brought the EPRDF to power in 1991 has informed the way the formal and informal institutions of government and military operate, creating a powerful military that is cohesive and a part of the wider Ethiopian vision for state development.

What makes Ethiopia, Rwanda and Uganda different to Senegal is that they have all experienced military rule at least once since gaining independence. Since rebels defeated their respective regimes, political leadership (who were former rebels) in Ethiopia (since 1991), Rwanda (since 1994) and Uganda (since 1986) have effectively built and maintained strong militaries that have not engaged in a coup, partly because the ex-rebels staffed the newly formed government and security apparatus. At the same time, ex-rebels in these new regimes created new political configurations in the government and military, along with new ideational and political programmes to broadly reinforce a new form of state-building. Not only did each of these regimes include former rebels in various departments and agencies, but they also integrated former regime officials as a way of facilitating reconciliation, unity and the process of nation- and state-building.

As suggested by Toft, rebel victories *can* lead to stability and 'durable outcomes'.[68] However, this is not a guarantee, as was the case in Mozambique, which witnessed a return to insurgent activity in 2013, despite peace having been reached in 1992.[69] Nonetheless, reconciliation efforts can help bring long-term stability after a civil war. In this context, Aisha Ahmad has identified the unrecognised state of Somaliland as being successful in the 1991 Somali civil war, attributing this to Somaliland's leaders engaging in reconciliation and allowing the losing rebels to join the government and military after victory / independence was declared.[70]

---

66 Author interviews with Ethiopian military personnel, 2017–18.
67 Author interviews with Ethiopian military personnel, August 2018.
68 Monica Duffy Toft, 'Ending civil wars: A case for rebel victory?', *International Security* 34:4 (2010), 7–36.
69 Friedrich Kaufmann and Winfried Borowczak, 'Mozambique's 'mini-war'', *Development and Cooperation*, May 2017, available at https://www.dandc.eu/en/article/ceasefire-fuels-hopes-peace.
70 Aisha Ahmad, 'The age of heroes', *TEDx Talks*, 9 June 2017, available at https://www.youtube.com/watch?v=oeg1tMnJWaM.

Gérard Prunier contends that Uganda and Rwanda developed strong armies due to geographical conditions and climate. He argues that due to the Tsetse fly not being a nuisance to human health and livestock, their region developed greater societal cohesion. From this, Prunier considers how Uganda could politically put together a strong military by solving inter-ethnic rivalries within the state. With the National Resistance Army (NRA) victory in 1986 – later renamed the Uganda People's Defence Force (UPDF) – Ugandan President Yoweri Museveni (a former NRA leader) effectively solved the tribal conflict between northerners and southerners. Museveni put less-favoured groups into specialised roles in the military and ones with coveted skills were given operational combat duties. Beyond this, he leveraged external economic and military assistance from the Soviets / Russians and the North Koreans (relations finally ceased in 2016), and also Western states in a pragmatically balanced fashion. Moreover, Museveni has been known to joke about keeping the UPDF busy with peacekeeping deployments abroad so that they cannot meddle in domestic politics.[71]

Nonetheless, it appears that the pathway to a strong military in Uganda has been a spirit of reconciliation between various ethnic / tribal groups, while providing a professional mission for the UPDF to train for and dedicate itself to. It should be no surprise that the UPDF was the first army to go on the offensive while conducting AU Mission in Somalia (AMISOM) peacekeeping operations in Mogadishu in February 2011. Moreover, with Uganda being the largest troop contributor to AU peacekeeping missions, this has helped Museveni rationalise his advocacy of the spirit of pan-Africanism, which is the ideological sinew that supports cohesion in his state by creating a sense of African nationalism in support of African states solving African problems. Such pan-Africanism also means that the UPDF is regularly used for domestic state-building activities to show how important the military is for creating a stronger and more developed Uganda. For instance, since 2006, the UPDF has held an annual 'Army Week' to positively engage the public by helping clear roadways, cleaning and updating medical facilities and improving conditions at some refugee camps.[72]

The Rwandan military and state has greatly advanced since the devastating genocide of 1994. In many ways, it is an army with a country,

---

71 Gérard Prunier, 'The Armies of the Great Lakes Countries', *PRISM* 6:4 (2017), 99–111.
72 Laura J. Perazzola, 'Civil-military operations in the post conflict environment: Northern Uganda case study', (US Naval Postgraduate School master's thesis, 2011).

where Paul Kagame led his rebel army in the overthrow of the genocidal Hutu regime (*Interhamwe*). Since taking over as president in 1994, he has sought to rebuild Rwanda through militaristic cohesion, relying heavily on US and British military assistance to achieve this. To this effect, President Kagame was responsible for the ideational concept of *Ingado*: espousing Rwandan nationalism to facilitate Hutu (re)integration socially and ideologically. Moreover, due to the neighbouring conflict in the Congo, Kagame found it useful to convert his Tutsi rebel army into a national Rwandan army (now known as the Rwanda Defence Force (RDF)), which integrated Hutus on equal terms with Tutsis.[73] This organisational strategy paid off, as it enabled Kagame to deploy thousands of RDF troops to Congo for his own reasons, while also participating in numerous international and regional peacekeeping operations. The RDF serves as a political vent because it acts as an alternate path to high-standing in the state, where Hutus can be co-equals in the Tutsi dominated regime. Additionally, the use of *Ingado* as a political programme informs the behaviour of the RDF, which is constantly engaged in domestic activities, ranging from agricultural support to corporate-industrial investment. This supports long-term Rwandan state-building efforts, while also permitting the RDF the ability to be militarily effective and professional.

Conclusion

The idea of a weak state with a strong military is a new arena for research, thus requiring further exploration. Moreover, with expectations that economic growth in Africa will continue to stagnate relative to the rest of the world, and that state weakness will be prevalent for the foreseeable future, it appears that stability in Africa will not occur due to democratisation. On the contrary, those weak African states that can create an effective armed force that contributes to societal cohesion and the overall paradigm of state-building, will be best placed to grow their economies – and eventually perhaps embark on democratic transition themselves.[74] Since armies typically play an important role in democratisation, a democratic

---

73 Marco Jowell, 'Cohesion through socialization: liberation, tradition and modernity in the forging of the Rwanda Defence Force (RDF)', *Journal of Eastern African Studies* 8:2 (2014), 278–93.
74 Jonathan Fisher and David M. Anderson, 'Authoritarianism and the securitization of development in Africa', *International Affairs* 91:1 (2015), 131–51.

transition too early in development or too rapidly pursued may do more harm than good, as an army might intervene out of self-preservation or because democratic reforms are too disruptive to informal institutions of power.[75] Similarly, as this chapter has shown, leadership in weak states must be resourceful and strategic in the way they utilise and fund their militaries. These leaders must also be careful in the way they absorb foreign military assistance, so that it is not treated purely as patronage and is put towards the overall nation- and state-building political programme.

Senegal put its military (and close ties to France) to good use in modernising society. Contemporary rebel-led states, such as Ethiopia, Uganda and Rwanda, effectively had to start from scratch in rebuilding the institutions of the state, especially the security apparatus, without a direct colonial linkage. Each of these states are representative of an *exercitu civitatis* ('army state)'. Their militaries are able to be effective and loyal to the state, contributing to societal and state cohesion. This is something that few armies can do in Africa without attempting a *coup d'état*, or, as seen throughout the Middle East, many of these coup-proofed armies collapse when fighting lightly armed insurgents.

For those wanting to address the weak state problem in Africa (and beyond), western policy-makers and scholars must start from the proposition that most weak states – if not all – will never govern 'like us' and that solutions for each state will require long-term partnerships.[76] This is equally tied to ensuring that leadership in weak states receives assistance in ways that contribute to the political strategy of long-term state-building, where an army can be a major component of such a venture if done properly. For instance, Rwanda's recovery is miraculous, especially in how integration of former enemies has worked in the long term to maintain societal and state cohesion, but improper integration of rebels – as seen in Burundi, the Central African Republic and the Congo – illustrates the dangers of not combining four vital ingredients: 'professionalization, socialization, welfare-provision and political education'.[77] Success in a weak state requires thoughtful political programmes that political, societal and military elites must commit to in the long term, foregoing the easy

---

75 Gabrielle Lynch and Gordon Crawford, 'Democratization in Africa 1990–2010: An assessment', *Democratization* 18:2 (2011), 275–310.
76 Melissa Annette Thomas, *Govern like us: US Expectations of Poor Countries* (New York: Columbia University Press, 2015).
77 Nina Wilén, 'From Foe to Friend? Army integration after war in Burundi, Rwanda and the Congo', *International Peacekeeping* 23:1 (2016), 79–106.

short-term pay-off of ethnic exclusion. If Western governments can better tailor assistance to host nation leaders to augment civil-military relations and improve societal cohesion, these weak states can develop strong armies that can begin resolving the insecurity issues that plague the continent, enabling African solutions to African problems.

References

Biddle, S. *Military Power: Explaining Victory and Defeat in Modern Battle*. Princeton NJ: Princeton University Press, 2010.

Braithwaite, J.M. and J.K. Sudduth. 'Military purges and the recurrence of civil conflict', *Research and Politics* 3:1, 2016.

Carroll, R.J. and B. Kenkel. 'Prediction, Proxies, and Power', *American Journal of Political Science*, 2019.

Diop, B. 'Civil-military relations in Senegal', in D.C. Blair (ed.). *Military Engagement: Influencing Armed Forces Worldwide to Support Democratic Transitions*. Washington DC: Brookings, 2013.

Dokken, K. *African Security Politics Redefined*. New York: Palgrave Macmillan, 2008.

Eizenstat, S.E., J.E. Porter and J.M. Weinstein. 'Rebuilding weak states', *Foreign Affairs* 84:1, 2005.

Fearon, J.D. and D.D. Laitin. 'Ethnicity, insurgency, and civil war', *American Political Science Review* 97:1, 2003.

Feaver, P.D. 'The civil-military problematique: Huntington, Janowitz, and the question of civilian control', *Armed Forces & Society* 23:2, 1996.

Fisher, J. and D.M. Anderson. 'Authoritarianism and the securitization of development in Africa', *International Affairs* 91:1, 2015.

Gaub, F. 'The Libyan armed forces between coup-proofing and repression', *Journal of Strategic Studies* 36:2, 2013.

Ghani, A. and C. Lockhart. *Fixing Failed States: A Framework for Rebuilding a Fractured World*. New York: Oxford University Press, 2009.

Howe, H.M. *Ambiguous Order: Military Forces in African States*. Boulder CO: Lynne Rienner, 2001.

Jackson, R.H. and C.G. Rosberg. 'Why Africa's weak states persist: The empirical and the juridical in statehood', *World Politics* 35:1, 1982.

Jowell, M. 'Cohesion through socialization: liberation, tradition and modernity in the forging of the Rwanda Defence Force (RDF)', *Journal of Eastern African Studies* 8:2, 2014.

Kalyvas, S.N. *The Logic of Violence in Civil War*. New York: Cambridge University Press, 2006.

Koonings, K. and D. Kruijt. *Political Armies: The Military and Nation Building in the Age of Democracy*. London: Zed Books, 2002.

Lyall, J. and I. Wilson 'Rage against the machines: Explaining outcomes in counterinsurgency wars', *International Organization* 63:1, 2009.

Lynch, G. and G. Crawford. 'Democratization in Africa 1990–2010: An assessment', *Democratization* 18:2, 2011.

Matisek, J. 'The crisis of American military assistance: strategic dithering and Fabergé Egg armies', *Defense & Security Analysis* 34:3, 2018.

Matisek, J. and W. Reno. 'Getting American Security Force Assistance Right: Political Context Matters', *Joint Force Quarterly* 92:1, 2019.

Migdal, J.S. *Strong Societies and Weak States: State-Society Relations and State Capabilities in the Third World.* Princeton NJ: Princeton University Press, 1988.

Perazzola, L.J. 'Civil-military operations in the post conflict environment: Northern Uganda case study'., US Naval Postgraduate School master's thesis, 2011.

Prunier, G. 'The Armies of the Great Lakes Countries', *PRISM* 6:4, 2017.

Reno, W. 'The politics of security assistance in the horn of Africa', *Defence Studies* 18:4, 2018.

Roessler, P. 'The enemy within: Personal rule, coups, and civil war in Africa', *World Politics* 63:2, 2011.

Talmadge, C. *The Dictator's Army: Battlefield Effectiveness in Authoritarian Regimes.* Ithaca NY: Cornell University Press, 2015.

Themnér, A. (ed). *Warlord Democrats in Africa: Ex-Military Leaders and Electoral Politics.* London: Zed Books, 2017.

Toft, M.D., 'Ending civil wars: A case for rebel victory?', *International Security* 34:4, 2010.

Weinstein, J.M. *Inside Rebellion: The Politics of Insurgent Violence.* New York: Cambridge University Press, 2006.

Wilén, N. 'From Foe to Friend? Army integration after war in Burundi, Rwanda and the Congo', *International Peacekeeping* 23:1, 2016.

Williams, R. 'Towards the Creation of an African Civil-Military Relations Tradition', *African Journal of Political Science /Revue Africaine de Science Politique* 3:1, 1998.

6

COLOMBIA AND THE PERSISTENCE OF 'FAILED STATE STIGMA' IN A PEACE SCENARIO

Saúl M. Rodríguez and Fabio Sánchez[1]

Introduction

There has been a widespread lack of understanding about the wider Colombian political situation. It seems as if, for the purposes of wider analytical examination, the state simply does not exist. Things began to change at the start of the 1980s, when it began to draw attention for its infamous cocaine cartels, the so-called 'war on drugs' and the increased irregular warfare against a number of guerrilla groups. These negative perceptions created what is referred to here as a 'failed state stigma', a distorted vision of the situation prompted by slanted media coverage and partial academic analysis. This situation was worsened when the consequences of the internal conflicts spilled over and began to affect the wider neighbourhood, whether via an increase in refugee movement at the borders, escalation of drug trafficking or the aerial fumigation of illegal crops.

To explore and further clarify the nature of said 'stigma' – and its consequences – this chapter proceeds as follows. In the first section, there is a historical analysis of the political context of Colombia, tracing the

1   The authors wish to thank Nathalia Alarcón for her research assistance. Translations from Spanish sources are the reponsibility of the authors.

development of violence, corruption and the 'war on drugs'. In particular, this section explores the range of diverse attempts within the existing literature to provide an accurate label for the condition of the Colombian state, whether it be 'co-opted', 'failed', 'failing' or just 'weak'. This will be followed by an examination of the most significant period of political turbulence – in the 1990s – when irregular actors attempted unsuccessfully to take control of the state, which ironically helps explain why Colombia has never been an absolute failed state. In the third section, an analysis of the political context that accompanied the peace agreement of 2016 is conducted. While, in some senses, this agreement serves as an example of the maturity, will and capacity of the Colombian state, nevertheless there persists a sense of political division – coupled with the unfortunate geographical truth that significant regions within Colombia effectively exist beyond the state's presence – that suggests there has been a partial collapse of the state.

Colombia: Once a Kind of Failed State?

Historically, power relations within Colombian society have been weak and problematic. Two centuries after independence, the state has not yet been able to fully and effectively control significant regions. This has assisted the rise of violent actors, who control strategic lands and resources, and helps explain the level and intensity of internal conflict, which has its roots in these problems. At the beginning of the 1960s, peasants took up arms illegally against the state under a variety of political flags, some of them under communist influence inspired by the Cuban Revolution. This problem was exacerbated by the penetration of ready and accessible drugs into all sectors of society: political, economic, the media and culturally, and in areas such as sport. In the 1980s, as a consequence, formerly ideologically committed guerrillas started to do business with the drug lords, transforming them into drug traffickers, leading to greater investment in heavy weapons and the establishment of their *de facto* political control of certain regions. Paramilitary forces emerged in response, with the use of effectively private armies to erode the guerrillas' control and with the wider civilian population trapped in the middle of such extreme violence. In some instances, these forces worked with the acquiescence of the state.[2]

---

2   Fernán González, Omar Gutiérrez, et al, *Conflicto y Territorio en el Oriente Colombiano* (Bogotá DC: CINEP, 2012), 145.

The increase in violence had other significant consequences beyond dividing the state. For instance, the impact on key services, such as medical assistance, the education system and public service coverage, was so bad that a 2005 Integral Development Index noted that, in some departments in southern Colombia, such as Amazonas, Guaviare, Guainía, Vaupés and Vichada, fewer than 40 per cent of the people were covered by these essential services, a situation made even less politically palatable when compared with the provision of a more effective service in the central zones.[3] This level of social inequality embodies the central problem of the Colombian state: a social fracture that has facilitated the growth of violence. In addition, a further five notable historical phases – which indicates the longevity of this problematic situation – can be identified as explanatory factors for the increase in violence: the level of corruption anchored within the system since colonial times; the level of rural violence in the late nineteenth century that culminates in a civil war called *Guerra de los Mil Días* (1899–2002); the period of partisan political violence between Liberals and Conservatives called *La Violencia* (1948–1958);[4] the prominence of the National Front (1958– 1974)[5] and the formation of guerrilla groups. These groups include Fuerzas Armadas Revolucionarias de Colombia (FARC), Ejército de Liberación Nacional (ELN) – both emerging in 1964 – the Ejército Popular de Liberación (EPL) in 1967 and paramilitary groups such as the Autodefensas Unidas de Colombia (AUC) and various other drug cartels during the 1980s and 1990s.

For all the potential differences in terms of motivation and mass, a similar pattern can be ultimately identified during each of the different stages of violence in Colombia, beginning with the role of economic and political elites, obsessed with controlling land, who seek to influence the poorer peasantry with competing political doctrines, both Liberal and Conservative, which clash bloodily in the absence of effective central state control. Some of the most productive lands were appropriated violently,[6] generating a widening social gap, where traditional political parties had links with landowners, marginalising

---

3   DNI, 'Medición y Análisis del Desempeño Integral de los Municipios – Departamento Nacional de Integración', *Dirección de Desarrollo Territorial Sostenible – DDTS*, 2005.
4   Arturo Alape, *El 9 de Abril, Asesinato de una Esperanza, Nueva Historia de Colombia* (Bogotá: Editorial Planeta, 1989).
5   This was a political agreement between the Liberal and Conservative political parties, that provided for their rotating power for 16 years (every four years they changed the presidential power). It was a way to end their political conflict, but at the same time it excluded other political alternatives and thus became in part the root of the contemporary conflict.
6   Luis Bértola and José A. Ocampo, *Desarrollo, Vaivenes y Desigualdad: una Historia Económica de América Latina desde la Independencia* (Madrid: Secretaría General Iberoamericana, 2010), 82.

the peasants further from the formal Colombian political system.[7] This helped fuel a rural crisis that pushed peasants to look for new lands in more remote regions, many of them beyond the apparent control of the state,[8] and where, once again, landowners had the monopoly and control of key aspects of social life.[9] Additionally, adding a new element to an identified historical cycle, in the early 1980s as noted, groups such as the FARC resorted to drug trafficking as a way of financing themselves,[10] which helped them buy more weapons to fight the state. At the same time, drug cartels, such as Cali's and Medellín's, the latter led by Pablo Escobar, waged vicious conflicts between themselves and with the state for control of key drug routes. Likewise, ranchers and mine owners created private military forces to control territories in the east of the country, a complex phenomenon of private ownership that eventually evaded even their control.[11] In essence, the result of these multifaceted developments is a state incapable of protecting its own people from a range of increasingly dangerous internal forces. In this respect, categorisations of the Colombian state as ranging from 'weak' to 'failed', seem accurate (if not politically sensitive), due to the state's inability to control private and profit-driven coercion and to exert its own formal sovereignty within all of its territory.

Searching for a Definition

Max Weber's vision of the state remains relevant and timely: 'a human community that, within a given territory (the territory is a distinctive

7   Fernán González, *Poder y Violencia en Colombia*, 4th ed (Bogotá: Odecofi-Cinep, 2016), 26.

8   Francisco Leal and Andrés Dávila, *Clientelismo: el Sistema Político y su Expresión Regional*, 3rd ed (Bogota: Universidad de los Andes, Facultad de Ciencias Sociales, Departamento de Ciencia Política, 2010).

9   Gonzalo Sánchez and Donny Meertens, *Bandoleros, Gamonales y Campesinos: el Caso de la Violencia en Colombia* (Bogotá: El Áncora, 1983); Germán Guzmán, Orlando Fals, and Eduardo Umaña, *La Violencia en Colombia: Estudio de un Proceso Social* (Bogotá: Carlos Valencia Editores, 1980); Malcom Deas and Fernando Gaitán, *Dos Ensayos Especulativos Sobre la Violencia en Colombia* (Bogotá: FONADE, Departamento Nacional de Planeación & Tercer Mundo Editores, 1995), available at http://www.unrisd.org/80256B3C005BCCF9/(httpPublications)/ 1F2B1C1475C25DCA80256B670065D2F1?OpenDocument&panel=relatedinformation; Gonzalo Sánchez, 'Violencia, Guerrillas y Estructuras Agrarias', in *Nueva Historia de Colombia. Historia Política 1946-1986* (Bogotá: Editorial Planeta, 1989), available at http://www. banrepcultural.org/blaavirtual/revistas/credencial/febrero1999/110laviolencia.htm.

10   Eduardo Pizarro, *Las Farc (1949-2011): de Guerrilla Campesina a Máquina de Guerra* (Bogotá: Grupo Editorial Norma, 2011).

11   Gustavo Duncan, *Los Señores de la Guerra: de Paramilitares, Mafiosos y Autodefensas en Colombia* (Bogotá: Planeta, 2006); Gustavo Duncan, *Más que Plata o Plomo: el Poder Político del Narcotráfico en Colombia y México* (Bogotá: Debate, 2014).

element), claims (successfully) for itself the monopoly on the legitimate use of physical violence'.[12] With adjustments made to the theoretical to take into account the practicalities of the contemporary world, Weber's definition points to a failed state as being a political entity that is unable to provide basic needs to its population: health, education, infrastructure, security, and thus to exercise sovereignty. However, it is asserted here that the Colombian case differs from classic examples of a 'complete' failed state – such as Somalia, South Sudan and the Democratic Republic of Congo (DRC) – with Colombia being more accurately recognised as a *limited* failure. In other words, it is a state where, in certain regions, its writ operates well, but – significantly in vast regions – the state simply does not exist as a factor in everyday life.

Even in the 1980s and 1990s, when the level of trafficking of illegal drugs destabilised the state to an even more significant extent, with attacks on towns (Mitú, 1988), military bases (Las Delicias, 1996), communication stations (Patascoy, 1997), pipelines (Caño Limón-Coveñas) and mass kidnappings in cities and across the countryside, it is difficult to argue the state had *completely* failed. In response, the government of Andres Pastrana (1988-2002) initiated an unfruitful peace negotiation with the FARC, giving them a demilitarised zone of 42,000 square kilometres where the armed group was able to strengthen itself and even legislate, although illegally, within its own 'zona de despeje', such as with the '001' expropriation of land, or the 10 per cent tax for people with over $1 million, or the '004' establishment of Justice Tribunals for politicians believed to be opposed to FARC. Outside this area, the Colombian state's political institutions remained operational and were never deprived of their constitutional powers at a general level. Such a situation, while bad, does not warrant the label and associated stigma of being referred to as a fully-failed state.[13]

Duncan and Oquist refer more accurately to a partial collapse of the state, predicated on 'the loss of control of the monopoly of violence by the legitimate security forces in large areas of the country',[14] which has facilitated the emergence of autonomous regional actors, such as warlords (paramilitary groups) or guerrillas (FARC, ELN, M-19) that fulfil the role of the state in political, administrative and judicial matters. In this way, these illegal actors have generated a para-state that has transformed local

---

12 Max Weber, *El Político y el Científico* (Madrid: Alianza Editorial, 1979).
13 See for example Charles T. Call, 'The Fallacy of the 'Failed State'', *Third World Quarterly* 29:8 (2008), 1491–1507.
14 Duncan, *Los Señores de la Guerra*, 42.

loyalties and resulted in the loss of overall state legitimacy.[15] Likewise, other authors have focused on the 'co-opted reconfiguration of the state' as a regulatory condition of regional relations, with local networks of power establishing ties with illegal groups to respond to the unmet needs of the rural population and the pressure of illegal groups to consolidate their political and economic power across the regions, effectively creating a limbo between legality and illegality.[16] Daniel Pécaut has coined the term 'precarious state' to describe a scenario in which the central institutions have failed to bridge the gaps between traditional and authoritarian social orders, thus generating a non-homogeneous democratic process within the country[17] which, according to Bejarano and Pizarro, indicates that the development of Colombia's democracy and statehood is uneven in itself, because the state cannot be equally present in every space. This facilitates the formation of quasi-independent republics, where violent actors provide basic services and offer parallel economic activities.[18]

In effect, the process of state formation in Colombia since independence has been characterised as a conflictive process regarding the acquisition and value of the land,[19] a struggle between the elites and the expansion of agricultural frontiers, where the state has not been able to respond positively, generating a 'domino effect' of violent actors that have supplanted the state in its functions. However, there has never been a total breakdown of political control and public order, even in the most critical moments of the 1990s, when some analysts seemed to conceive of Colombia as a kind of declining state, more than a failed one.[20]

The 1990s: A Lost Decade of Fatal Violence

The beginning of the 1990s represented a complex context in which various violent actors attempted to establish their own agendas in defiance of the

---

15 Paul Oquist, *Violencia, Conflicto y Política en Colombia* (Bogotá: Instituto de Estudios Colombianos, 1978), 277.
16 Luís Jorge Garay et al (eds), *La Captura y Reconfiguración Cooptada del Estado en Colombia* (Bogotá: Grupo Método – Transparencia por Colombia – Fundación AVINA, 2008), 80.
17 *Guerra Contra la Sociedad* (Bogotá: Espasa, 2001), 118.
18 *From 'Restricted' to 'Besieged': the Changing Nature of the Limits to Democracy in Colombia* (Notre Dame IN: Helen Kellogg Institute for International Studies, 2002), 14.
19 Salomón Kalmanovitz, *Nueva Historia Económica de Colombia* (Bogotá: Fundación Universidad de Bogotá Jorge Tadeo Lozano / Aguilar, Altea, Taurus, Alfaguara, S.A., 2010), 268.
20 Call, 'The Fallacy of the 'Failed State''.

Colombian state: guerrillas (FARC, ELN, EPL), paramilitaries and the Medellín and Cali cartels. The significance of such violent actions could be underestimated, due to the heterogeneity and all-too-regular distribution within the national territory, with the people becoming accustomed to such levels of violence and the presence of illegal groups everywhere.[21] However, in reality, this period was critical, with violence escalating at an alarming pace, reaching its peak in 1991, when 'the homicide rate tripled in size from 25 homicides per 100,000 inhabitants in 1974 to 79 in 1991'.[22] Additionally, more than 500,000 people were displaced by the violence and the number of kidnappings ranged between 1,000 and 1,717 people per year.[23] The actions of the Medellín Cartel, led by Pablo Escobar, killed at least 5,500 people between 1989 and 1993 alone.[24] Among their most violent acts, they took down an Avianca jet plane in 1989 causing 110 deaths, and put three car bombs in the city of Bogotá in 1993 leaving at least 332 wounded and 33 dead. [25] Coupled with the death toll attributed to the Cali Cartel in Medellín in 1995 that left 20 dead and 99 wounded, it is significant that such internal actors were able to act with such seeming impunity, their power increasing at the expense of the state.[26]

Paradoxically – and perhaps contributing to a relative lack of societal panic at the escalation of violence – moves were made toward some level of state modernisation, helped by the rise of a new generation of politicians that aimed to more fully connect Colombia with the increasingly globalising wider world. Given the economic opening, a new constitution was passed that did not permit extradition, and peace negotiations were initiated with both the M-19 and EPL guerrilla groups. The new constitution deserves special attention, as much for what was left out in order to gain acquiescence from the guerrilla groups, such as tackling forced displacement, the role of corrupt politicians and the existence of paramilitaries (some of which had ties to the military), as for what was actually agreed. Therefore, the political and economic reforms at this time were highly short-sighted and, by default,

---

21 Pécaut, *Guerra Contra la Sociedad*, 189.
22 Catalina Bello, 'La Violencia en Colombia: Análisis Histórico del Homicidio en la Segunda Mitad del Siglo xx', *Revista Criminalidad* 50:1 (2008), 73-84, available at http://www.scielo.org.co/pdf/crim/v50n1/v50n1a05.pdf.
23 Pécaut, *Guerra Contra la Sociedad*, 182.
24 Semana, 'Cifras de Atentados y Víctimas de Escobar' (Bogotá: 2013), available at https://www.semana.com/nacion/articulo/cifras-de-atentados-victimas-de-escobar/365633-3.
25 *El Tiempo*, 'Historia de Otras Bombas – Archivo Digital de Noticias de Colombia y el Mundo Desde 1990', (1999), available at https://www.eltiempo.com/archivo/documento/MAM-949482.
26 *El Tiempo*, 'Bomba en Medellín deja 20 Muertos y 99 Heridos', (1995), available at https://www.eltiempo.com/archivo/documento/MAM-343371.

strengthened both drug trafficking and illegal armed groups. In terms of policy reforms, a decentralisation process was initiated that provided for greater autonomy for local and regional authorities to administer the territory. This only reinforced the existing weakness of central government power, creating further possibilities for insurgent groups to become more involved and influence central policy through networks of corruption and coercion, on some occasions coercing voters during elections.[27]

Having established as a response to weakening centralised power a further set of channels for local actors, the state's wider economic reforms and neoliberal opening – including an attempted Agrarian Reform – created further insecurity in rural communities and provided more discontent for newly empowered local actors to feed on. By enabling the creation or recreation of large estates, the law created a kind of *agrarian counter-reform*[28] in 1993, because government initiatives had created greater conditions of impoverishment and marginalisation within the farming community, leading many to turn to coca production as an alternative for survival as, in most cases, it gave them higher revenues per kilogram than traditional crops.[29] Also, the irregular groups took advantage of land titling and appropriations developed by the government to acquire many of these properties at a low cost. An alarming sign of this was that at least 11 per cent of rural properties were acquired by drug traffickers at this time.[30] As a consequence, a complex network of cultivation, extraction, processing and transportation of illicit substances led to Colombia becoming the largest producer of coca in the Andean region by 2002, accounting for 59 per cent of all illicit crops. This remains a significant problem that has continued to increase, despite different strategies to control or destroy it by the state, such as manual eradication, aerial fumigation and alternative crops, such as the palm heart producers in Putumayo. At the time of writing in 2019, according to the Drugs Observatory from the Justice Ministry, there are still 171,494 hectares of land dedicated to the development of coca crops.[31]

---

27 Mary Roldán, 'End of Discussion: Violence, Participatory Democracy, and the Limits of Dissent in Colombia', in Enrique Desmond and Daniel Goldstein (eds), *Violent Democracies in Latin America* (Durham NC: Duke University Press, 2010), 64.

28 González, *Poder y Violencia en Colombia*, 412.

29 Ricardo Rocha, 'La Riqueza del Narcotráfico y la Desigualdad en Colombia, 1976–2012', *Revista Criminalidad* 56:2 (2014), 273–90, available at https://ideas.repec.org/p/col/000118/011886.html.

30 Duncan, *Los Señores de la Guerra*, 287.

31 ODC, 'Estadísticas Nacionales', 2019, available at http://www.odc.gov.co/sidco/perfiles/estadisticas-nacionales.

Some drug resources became war supplies that permitted illicit groups to acquire more and better weapons, making it easier for them to increase their presence in Colombian municipalities. Thus, groups such as the ELN expanded considerably, to as many as 4,000 combatants,[32] with the FARC active on 63 rural and four urban fronts[33] and with the paramilitary groups also creating semi-regular armies, which in April 1997 coalesced into the AUC, under the leadership of Carlos Castaño.[34] The AUC fought both the state and other guerrillas and attacked arbitrarily the civilian population, accusing them of sympathising with their enemies. This in turn generated a further humanitarian tragedy in the state, which led to an increase in the number of Internally Displaced Persons (IDP) and areas seeded with anti-personnel mines, with Colombia holding the unwanted record of being the second most mined state in the world after Afghanistan.[35]

Building on an already chequered history, such developments during the 1990s created an explosive cocktail for Colombian society. The country suffered a reversal of values, with the imposition of the rule of a new social class advanced by drug money. Between 1990 and 1993 there was an increase in the homicide rates in urban areas of the country's main cities – Medellín, Cali and Bogotá – with Medellín recognised as the most violent city in the world at this time, surpassing homicide rates in areas of armed conflict or civil war, such as Sierra Leone and Liberia.[36] Even with the decline of the Medellín and Cali cartels, the drug business simply fell into the hands of smaller and less-prominent groups, as well as the FARC, which, while attempting to seize power through force of arms, consolidated itself as another drug cartel, controlling routes, extorting small producers and increasing illegal actions against the civilian population. The statistics from this period are alarming:

- From an analysis of the 342 municipalities with the highest rates of homicide and/or kidnapping and/or high-intensity armed conflict

32 Camilo Echandía, 'Auge y Declive del Ejército de Liberación Nacional (ELN): Análisis de la Evolución Militar y Territorial de Cara a la Negociación' (Bogotá: Fundación Ideas Para la Paz 2013), 5–22, available at http://cdn.ideaspaz.org/media/website/document/529debc8a48fa.pdf.

33 María Alejandra Vélez, 'FARC-ELN: Evolución y Expansión Territorial', *Revista Desarrollo y Sociedad* 47 (2001), 167.

34 Duncan, *Los Señores de la Guerra*, 340.

35 RCN, 'En un País Minado no es Posible una Paz Completa: Santos Noticias RCN', 2016, available at https://noticias.canalrcn.com/nacional-pais/un-pais-minado-no-posible-una-paz-completa-santos.

36 The so-called 'Pistola Plan' was especially notorious. Pablo Escobar paid more than $200 for each police officer murdered in the city of Medellín.

between 1993 and 1995, guerrillas were present in 284 (83 per cent) of them[37]

- The number of kidnapping claims per capita tripled in the country. Attempts against personal freedom reported by households in victimisation surveys increased nearly tenfold between 1985 and 1995. Finally, the reporting of terrorist acts increased by 86 per cent.[38]

These statistics indicate the high impact that the various manifestations of violence – particularly guerrilla violence – had on Colombia during the 1990s. Arguably worse in terms of the 'failed state stigma', the scale and intensity of such conflict had drawn the attention of external actors, in the US, Europe and other international observers. In effect, Colombia – for good or ill – had ceased to be another forgotten conflict in the Americas, overshadowed by the military dictatorships of the Southern Cone or the revolutions in Cuba and Nicaragua. Colombia had become a problematic country as an exporter of drugs, full of radical guerrillas and facing an internal humanitarian tragedy, the latter producing a spillover of weapons, guerrilla camps and refugees into neighbouring states. Therefore, at the end of that unfortunate decade, Colombian President Andres Pastrana and US President Bill Clinton created the 'Plan Colombia',[39] a military and social cooperation programme that intended to reassert territorial control, although without solving the main social and political problems related to the regions, such as the weak social, educational and health programmes.

Democratic Security: Against the 'Failed State Stigma'

At the turn of the new millennium, Colombia was in a dire state. Guerrilla groups such as FARC, ELN and the EPL had reached a position of significant power across a number of regions. Likewise, the brutal crimes committed by right-wing paramilitary groups against members of the civil population accused of being guerrilla collaborators gave an impression that all was lost for the country. Perceptions matter and, in the US, Colombia's main

---

37 Echandía, 'Expansión Territorial de La Guerrilla Colombiana: Geografía, Economía y Violencia', 139.

38 Mauricio Rubio, 'Violencia y Conflicto en los Noventa', *Coyuntura Social*, 2000, 172, available at https://www.repository.fedesarrollo.org.co/handle/11445/1772.

39 María Catalina Monroy and Fabio Sánchez, 'Foreign Policy Analysis and the Making of Plan Colombia', *Global Society* 31:2 (2017), 245–71.

partner in the 'war on drugs', politicians and scholars began to more regularly explore and publicise the possibility that the Colombian state could collapse.[40] Such external disappointment and disapproval in and of itself adversely affected the self-confidence of the very local political elites that were necessary to combat such violent trends, those who had previously been proud to belong to one of the oldest electoral democracies and, relatively, one of the most stable economies in Latin America.[41] When combined with the fallout from the failed peace negotiation with FARC, public opinion was left in a contradictory state, showing both wrath and fear about the deceitful position of the guerrilla group. In the short term, the first significant political consequence of this legacy was the successful elevation of Alvaro Uribe Vélez to the Colombian presidency in 2002.

Uribe's advantage was that, as an outsider, he did not belong to the Colombian traditional elites, who were believed both internally and externally to have failed to combat the various violent trends that were weakening the state. Uribe's political conception centred on the idea that Colombia has had a long-lasting tradition of civilian governments that had lost the capacity to control its territory over the years, partly due to negligence and also to the insufficient demonstration of authority. Uribe's central policy was known as 'Democratic Security' (DS), a counterinsurgency and counterterrorist programme that deployed all military capabilities under civilian management to retake control, with a particular focus on the FARC's bulwarks in the southern and eastern regions of the state. The central characteristics of this policy were to strengthen the state via improved military and judicial capabilities, enhance the intelligence and counterintelligence activities of the state and, as a consequence, build confidence within the wider citizenry.[42] Militarily speaking, the novel aspects of Uribe's plan – which built upon indigenous armed forces enhanced by US support as part of 'Plan Colombia' – was his leadership of the armed forces on the ground. To increase military effectiveness in all stages of the Colombian armed forces, there was an updated recruiting programme of peasant soldiers and regular troops and new economic resources based on taxes collected from wider sectors of society. From as

---

40 See Phillip McLean, 'Colombia: Failed, Failing, or Just Weak?', *The Washington Quarterly* 25:3 (2002), 123–34.

41 Andrés Pastrana, 'Texto de la Alocución del Presidente Andrés Pastrana', *El Tiempo*, 2002, available at http://www.latinamericanstudies.org/colombia/alocucion.htm.

42 Presidencia de la República; Ministerio de Defensa Nacional, *Política de Defensa y Seguridad Democrática* (Bogotá: Presidencia de la República, 2003), available at https://www.oas.org/csh/spanish/documentos/colombia.pdf.

early as 2003, the combination of military force and police action greatly reduced the guerrilla military actions in several regions, retook towns that had been beyond state control for many years and pushed guerrilla groups to the periphery of the state.[43] Progressively, and thanks to government propaganda that showcased the achievements of this policy to regain geographic control, both central audiences – the local population and key international investors – recovered their trust in the state's capacity to act. In this respect, Foreign Direct Investment (FDI) increased from $2.1bn in 2002 to $13.2bn in 2012.[44]

However, these achievements were not uniformly accepted. There were accusations of human rights violations and unfair prosecution of political opponents carried out by state forces. Perhaps the most notorious case was the so-called *falsos positivos*, a series of extrajudicial killings committed by agents of the Colombian military forces. Many army members still await judicial investigation at the time of writing.[45] These accusations damaged the positive image that the government was trying to project, to shift the 'failed state stigma', but, thanks to continued popular confidence in Uribe and the support of the most influential Colombian mass media, they were rapidly overshadowed, despite the dubious tactics adopted. While some may have considered that the ends justified the means in terms of restoring much-needed levels of control by the state, there remains an issue over how such revived state capacity was deployed. Alongside this hard-line approach, the Uribe government also implemented a peace process with the AUC, which had been accused of more than 60 per cent of the total crimes against the Colombian civilian population during the last years of the internal armed conflict.[46] This was a measure ultimately worth taking, as these groups officially demobilised after agreement was reached in the mid-2000s, which also improved confidence abroad, particularly in the US about the domestic state-building process led by Uribe. Regardless of such efforts, there was still some lower-level activity by such groups acting

---

43 Andrés Dávila, Miguel Gomis, and Gustavo Salazar, 'El Conflicto en Contexto 1998–2014', in *El Conflicto en Contexto* (Bogotá: Javeriana, 2016), 79–86.
44 See Park Madison Partners, 'Colombia's Rise: a Primer for International Investors', Park Madison Partners, 2013, available at http://www.parkmadisonpartners.com/documents/FG/parkmadison/news/15518_Colombia's_Rise_A_Primer_for_International_Investors_January_2013.pdf.
45 Javier Castrillón and René Guerra, 'A Deep Influence: United States-Colombia Bilateral Relations and Security Sector Reform, 1994–2002', *OPERA* 20 (2017), 35.
46 Centro Nacional de Memoria Histórica, 'Estadísticas del Conflicto Armado en Colombia', 2012, available at http://www.centrodememoriahistorica.gov.co/micrositios/informeGeneral/estadisticas.html.

locally, committing crimes related to drug trafficking and killing human rights defenders and opposition leaders.[47]

Even with such continued – and, it might be argued, normalised – violence, for the Uribe government, the military priority remained the FARC. In this respect, in the last years of his term, Uribe's government increased the intensity of action against this illegal group, with the killing of top-ranking leaders as a result of air raids and intelligence operations executed by Colombian military forces. Doubtless, the most important action was the bombardment of the FARC leader, Raul Reyes, in 2008, on the Colombian-Ecuadorian border. In some way this proved that the Colombian state was now able to significantly degrade any illegal force that continued to defy state institutions.[48] While successful internally, given the location of such actions, these generated diplomatic problems within the neighbourhood, with Colombia criticised for the spillover of its internal conflict beyond its borders.[49] This actually improved the credibility of the case that Colombian officials were making against the 'failed state stigma', although labelling it as such permitted more radical measures to be taken as necessary to save the country from such a continued threat.[50] While it may have been historically more relevant and of continued practical utility in some ways, the label of 'failed state' – which much of Colombian academia had already rejected – was during the 2000s of lessened validity, given that the FARC was weakened significantly during this period. As part of the wider, zero-sum game between state forces and armed non-state actors, there was a noted shift, with the presence of civil and military authorities increasing from 300 municipalities in 2002 to 1,098 in 2006: a positive result regarding both the consolidation of state power and also perceptions.[51]

---

47 Congressional Research Service, 'Colombia: Background and U.S. Relations' (Washington DC: CRS, 2018), 14-15, available at https://fas.org/sgp/crs/row/R43813.pdf.

48 Simón Romero, 'Colombian Forces kill Senior Guerrilla Commander, Official says', *New York Times*, 2 March 2008, available at https://www.nytimes.com/2008/03/02/world/americas/02farc.html.

49 Fabio Sánchez, 'Cooperation and Discord in South America in the Twenty-first Century: the Consequences of the Colombia-United States Military Agreement of 2009', in Bahram M. Rajaee and Mark J. Miller (eds), *National Security under the Obama Administration* (New York: Palgrave Macmillan, 2012), 159-76.

50 *El Mundo*, 'Un Presidente Obsesionado con Acabar con las Farc', 26 February 2010, available at https://www.elmundo.es/america/2010/02/27/colombia/1267229275.html.

51 Samir Elhawary, 'Security for Whom? Stabilisation and Civilian Protection in Colombia', *Disasters* 34 (2010), S388–405.

FARC's Threat, the Peace Agreement and the Partial Collapse of the State

Thanks to this, Juan Manuel Santos, Uribe's former minister of defence, was elected as his successor for the 2010–14 presidential term, to continue with 'Democratic Security' and the hard-line perspective against FARC. This proved initially successful, with the deaths of two of the most important leaders of FARC; Mono Jojoy in 2010 and Alfonso Caño the following year. They represented the military and political wing of FARC respectively. With this military success, many people expected that Santos would intensify operations against the group in a large-scale military operation, especially as the remainder of the guerrilla forces were still active and spread across the country. FARC also showed signs of adaptation to government tactics, including using smaller military units, landmines and snipers to combat the aerial advantage of state forces.[52] Perhaps because of this continued if lower-level threat, Santos unexpectedly announced in 2012 the beginning of a peace negotiation with FARC.

While seemingly surprising, given the initial continuity with his predecessor, this decision was in line with Santos' longer-stated understanding that a military victory was ultimately impossible, with a secure peace agreement a more stable solution to avoid more bloodshed between Colombians. He granted political status to FARC by recognising the reality that an internal conflict in Colombia continued to exist and then, at a greater political risk, beginning in 2011, he carried out secret negotiations with the group, supported by states such as Norway and Cuba. The Havana talks between 2012 and 2016 discussed some of the more critical issues regarding the nature of the domestic conflict, including drug trafficking, land reform, future political participation of guerrillas, rights of victims and the steps needed to ensure the implementation of the agreement.[53] Doubtless, this was a unique opportunity to negotiate with the most dangerous actor within the state, which had attempted to overthrow established political power for almost 50 years. Despite opposition, notably from former president Uribe and his party who believed the process would simply reward FARC and give them the political room for manoeuvre to

---

52 Andrés Dávila, Gustavo Salazar, and Alexander González, *El Conflicto en Contexto: un Análisis en Cinco Regiones Colombianas, 1998–2014* (Bogotá: Pontificia Universidad Javeriana, 2016).
53 Dávila, Salazar, and González.

defeat the state in the future,[54] in late 2016 the agreement was signed after the approval of the Colombian Congress.[55]

In certain respects, the military campaign and the peace agreement with FARC serve as proof that the shadow of the 'failed state stigma' is over, because the Colombian state was able to conduct both the latter stages of the war and then the peace in a proper way, despite multiple negative claims to the contrary. However, the overt focus on FARC has merely distracted the state from other equally pressing concerns, not least that there remains a vacuum of state control in several regions.[56] A number of right-wing groups and FARC dissidents have taken control of some of FARC's former territories, thereby repeating the historical cycle and once again bypassing the state. Secondly, corruption has become an even more notorious issue, with damaging reports of local political forces in league with a range of illegal groups, which further weakens the credibility of the state in these regions. Thirdly, in some cases, the state has been unable to guarantee the lives of former FARC combatants, underlining perhaps both the lack of willingness and capacity to be seen to fulfil some of the more controversial aspects of the peace agreement.[57] Fourthly, the promotion of human rights remains a dangerous activity in Colombia, with more than 400 activists killed by illegal forces in 2018 alone.[58] In other words, while the wider stigma of state failure may have been removed from the political and public arena, both within Colombia and externally, the reality of the state's continued vulnerability and fragility very much remains in certain more remote zones of the national territory.

---

54 Semana, 'Para Uribe, el País no Debe Aceptar el Acuerdo de Víctimas', (Bogotá: 2015), available at https://www.semana.com/nacion/articulo/proceso-de-paz-la-posicion-de-alvaro-uribe-frente-el-acuerdo-de-victimas/453787-3.

55 Caracol TV.com, 'Santos y 'Timochenko' Firman el Acuerdo de Paz Definitivo', *Noticias Caracol* 2016, available at https://noticias.caracoltv.com/acuerdo-final/delegados-de-gobierno-y-farc-se-reunen-en-bogota-para-firma-del-nuevo-acuerdo.

56 We follow some ideas proposed by Paul Oquist in this context. See Oquist, *Violencia, Conflicto y Política en Colombia.*

57 *El Espectador*, 'Las 9 Preocupaciones de la Onu Sobre la Implementación del Acuerdo de Paz' (Bogotá: 2019), available at https://colombia2020.elespectador.com/pais/las-9-preocupaciones-de-la-onu-sobre-la-implementacion-del-acuerdo-de-paz; Semana, 'Asesinan a dos Desmovilizados de las Farc', 2018, available at https://www.semana.com/nacion/articulo/asesinan-a-dos-desmovilizados-de-las-farc/553953.

58 France24, 'En 2018 empeoró la situación de los líderes sociales en Colombia', 17 December 2018, available at https://www.france24.com/es/20181214-lideres-sociales-colombia-asesinatos-2018.

Conclusion

Colombia represents a somewhat unusual case of survival amid relentless cycles of violence. This chapter has outlined briefly how the state has been marked by a constant political divide, on class, geographical and ideological grounds, which has ultimately given rise to guerrilla groups such as FARC. While the prominence of the Marxist armed struggle faded with the introduction of drugs to their agenda – leaving them, by the 1980s as narco-terrorist clans, as was noted at that time by US Ambassador to Colombia, Lewis Tambs[59] – the impact of the violence on the ordinary citizens of the state remained effectively the same.

At that time, the US-inspired 'Plan Colombia' seemed to be the solution to deal with at least the first level of violence, in terms of the actions of drug traffickers, guerrillas and paramilitaries. However, the roots of the problem lay in the continued state vacuum, with drug production – as in states such as Afghanistan – remaining a viable alternative for large, marginalised sectors of Colombian society. Additionally, even the more hard-line administration of Alvaro Uribe succeeded in only weakening violent actors, rather than the complete elimination of all non-state armed groups, assumedly a necessary precondition for the creation of a Weberian monopoly of force by the central state authorities. As well as failure at that level, some of the tactics adopted – which may have demonstrated the capacity but not the credibility of the state – led to accusations of human rights abuses, which may constitute another level of failure.[60] This was followed, perhaps as a consequence of the failure of force to restore complete state control, by the Santos peace process, although this, in and of itself, has not resolved the underlying tensions between the state and FARC and has arguably led to a greater level of political polarisation over whether such an agreement is justified and necessary.

Nevertheless, with the peace treaty signed in 2016, the country seemed to enter a new era of pacification. However, given the continued underlying social problems affecting Colombia – such as the lack of nationally effective healthcare, education, housing and security – at most the state has tackled the symptoms, not the underlying causes of its own fragility. In addition, the continued prominence given to a high degree

---

59 Francisco Coy, 'Injerencia Creciente y Desnarcotización Fallida: las Relaciones Colombia-Estados Unidos del Final de la Guerra Fría', *Desafíos* 9 (2003), 165–92.

60 BBC Mundo, 'Difícil Camino para Uribe en ee.uu', 2017, available at http://news.bbc.co.uk/hi/spanish/latin_america/newsid_6590000/6590521.stm.

of corruption at various political levels has led to accusations that the war against FARC itself served merely as a smokescreen for actors who for years have been stealing from the Colombian treasury. Paradoxically, while the state is at a crossroads, divided anew between the critics and revisionists of the peace process, led by Alvaro Uribe, and its supporters, Colombia has made strides externally. This led to the international irony that a state that cannot provide sufficient security and good governance to all its people became a new 'global partner' of NATO[61] and a member of the Organisation for Economic Cooperation and Development (OECD).[62] It remains to be seen whether international perceptions of the current Duque administration will remain more positive, as it seeks to avoid falling once again into another wave of cyclical violence and therefore finally overcomes the stigma of being viewed, both internally and as importantly externally, as a partially collapsed or failed state.

References

Bello, C. 'La Violencia en Colombia: Análisis Histórico del Homicidio en la Segunda Mitad del Siglo xx', *Revista Criminalidad* 50:1, 2008.

Bértola, L. and J.A. Ocampo. *Desarrollo, Vaivenes y Desigualdad: una Historia Económica de América Latina desde la Independencia*. Madrid: Secretaría General Iberoamericana, 2010.

Call, C.T. 'The Fallacy of the 'Failed State'', *Third World Quarterly* 29:8, 2008.

Castrillón, J. and R. Guerra. 'A Deep Influence: United States–Colombia Bilateral Relations and Security Sector Reform, 1994-2002', *OPERA* 20, 2017.

Coy, F. 'Injerencia Creciente y Desnarcotización Fallida: las Relaciones Colombia-Estados Unidos del Final de la Guerra Fría', *Desafíos* 9, 2003.

Dávila, A., M. Gomis and G. Salazar. 'El Conflicto en Contexto 1998–2014', *El Conflicto en Contexto*. Bogotá: Javeriana, 2016.

Deas, M. and F. Gaitán. *Dos Ensayos Especulativos Sobre la Violencia en Colombia*. Bogotá: FONADE, Departamento Nacional de Planeación & Tercer Mundo Editores, 1995.

Duncan, G. *Los Señores de la Guerra: de Paramilitares, Mafiosos y Autodefensas en Colombia*. Bogotá: Planeta, 2006.

Duncan, G. *Más que Plata o Plomo: el Poder Político del Narcotráfico en Colombia y México*. Bogotá: Debate, 2014.

---

61 ABC, 'Colombia, the first country in Latin America that Enters NATO as 'Global Partner'', 2018, available at https://www.abc.es/internacional/abci-colombia-primer-pais-america-latina-ingresa-otan-201805271724_noticia.html.

62 'Colombia and the OECD' (OECD, 2018), available at https://www.oecd.org/centrodemexico/laocde/colombia-y-la-ocde.htm.

Echandía, C. 'Auge y Declive del Ejército de Liberación Nacional (ELN): Análisis de la Evolución Militar y Territorial de Cara a la Negociación'.,Bogotá: Fundación Ideas Para la Paz, 2013.

Elhawary, S. 'Security for Whom? Stabilisation and Civilian Protection in Colombia', *Disasters* 34, 2010.

Everton, E. et al. 'Brokers and Key Players in the Internationalization of the Farc', *Studies in Conflict & Terrorism* 36:6, 2013.

Garay, L.J. et al. (eds). *La Captura y Reconfiguración Cooptada del Estado en Colombia.* Bogotá: Grupo Método – Transparencia por Colombia – Fundación AVINA, 2008.

González, F. and O. Gutiérrez. *Conflicto y Territorio en el Oriente Colombiano.* Bogotá: CINEP, 2012.

González, F. *Poder y Violencia en Colombia* (4th ed.).,Bogotá: Odecofi–Cinep, 2016.

Guzmán, G., O. Fals and E. Umaña. *La Violencia en Colombia. Estudio de un Proceso Social.* Bogotá: Carlos Valencia Editores, 1980.

Leal, F. and A. Dávila. *Clientelismo: el Sistema Político y su Expresión Regional* (3rd ed.).,Bogotá: Universidad de los Andes, Facultad de Ciencias Sociales, Departamento de Ciencia Política, 2010.

McLean, P. 'Colombia: Failed, Failing, or Just Weak?', *The Washington Quarterly* 25:3, 2002.

Monroy, M.C. and F. Sánchez. 'Foreign Policy Analysis and the Making of Plan Colombia', *Global Society* 31:2, 2017.

Oquist, P. *Violencia, Conflicto y Política en Colombia.* Bogotá: Instituto de Estudios Colombianos, 1978.

Pizarro, E. *Las Farc (1949-2011): de Guerrilla Campesina a Máquina de Guerra.* Bogotá: Grupo Editorial Norma, 2011.

Pizarro, E. and A.M. Bejarano. 'Colombia: a Failed State?', *ReVista: Harvard Review of Latin America* 2:1, 2003.

Rocha, R. 'La Riqueza del Narcotráfico y la Desigualdad en Colombia, 1976-2012', *Revista Criminalidad* 56:2, 2014.

Roldán, M. 'End of Discussion: Violence, Participatory Democracy, and the Limits of Dissent in Colombia', in E. Desmond,and G. Goldstein (eds), *Violent Democracies in Latin America.* Durham NC: Duke University Press, 2010.

Rubio, M. 'Violencia y Conflicto en los Noventa', *Coyuntura Social,* 2000.

Sánchez, F. 'Cooperation and Discord in South America in the Twenty-first Century: the Consequences of the Colombia–United States Military Agreement of 2009', in B.M. Rajaee, and M.J. Miller (eds), *National Security under the Obama Administration.* New York: Palgrave Macmillan, 2012.

Sánchez, G. and D. Meertens. *Bandoleros, Gamonales y Campesinos: el Caso de la Violencia en Colombia.* Bogotá: El Áncora, 1983.

Sánchez, G. 'Violencia, Guerrillas y Estructuras Agrarias', in *Nueva Historia de Colombia: Historia Política 1946–1986.* Bogotá: Editorial Planeta, 1989.

Vélez, M.A. 'FARC-ELN: Evolución y Expansión Territorial', *Revista Desarrollo y Sociedad* 47, 2001.

7

FOREIGN INTERVENTION AND THE PROCESS OF STATE FAILURE

A Case Study of Libya, 2011

Islam Goher

Introduction

Over the last three decades, the phenomenon of 'state failure' has become a defining pattern for many Arab countries. The role of foreign intervention was a crucial reason explaining several cases of such failure.[1] One of the most outstanding cases is Libya. In 2011, Libya witnessed rapid political unrest, followed by a humanitarian emergency. As a result, the United Nations Security Council (UNSC) authorised international intervention. The initial verdicts that the declared humanitarian intervention was a resounding success have subsequently been substituted by pessimistic

---

1 Iqbal and Starr have highlighted the important role of international intervention in the phenomenon of state failure. They argue that 'State failure, like any other phenomenon ... its onset, termination, duration, recurrence, and consequences are bound up with ... the involvement of the international community', and hence they reiterate that 'assessments of international involvement before collapse, and intervention after collapse, would yield valuable insights into the occurrence of this phenomenon'. See Zaryab Iqbal & Harvey Starr, *State failure in the modern world* (Stanford CA: Stanford University Press, 2015), 124–5.

conclusions that Libya is becoming a failed state.[2] This has been reflected in many indexes that measure state weakness and fragility. Libya's ranking, according to the Fund for Peace's Fragile States Index, for example, has shifted in the aftermath of its revolution from 111th most fragile in 2011 to 25th in 2018. According to the Corruption Perceptions Index, its ranking in this area moved from 146th in 2010 to 171st in 2017.[3]

This chapter investigates the relationship between foreign intervention and state failure. It particularly addresses the question of how humanitarian intervention, which is supposed to protect human rights, could lead to a level of state failure that, paradoxically, endangers people's lives. In addressing this question, the chapter will focus on the Libyan case study in 2011 and will be divided into five parts. The first develops a theoretical framework for the mechanism through which foreign intervention could lead to state failure. This will then be applied to the Libyan case study to more thoroughly assess the mechanism in an empirical setting. The second part examines the internal determinants for intervention within Libya, complemented by the third, which examines external determinants. The focus of the fourth part will be on the mechanisms of actual interventions in Libya in 2011, before the chapter concludes with an assessment of the impact of intervention on existing and developing political, security and economic levels.

Foreign Intervention and State Failure

Within the state failure literature, there is widespread agreement that weak, failed and collapsed states are exposed to varying degrees of intervention. However, scholars and analysts disagree on the role of external intervention. There are two main approaches presented in this regard. The first views intervention as a *consequence* of state failure, because such failure has such significant repercussions for regional and international security that external intervention is required either to control or rebuild such states.[4]

---

2   See, for example, C. Hobson, 'Responding to Failure: The Responsibility to Protect after Libya', *Millennium: Journal of International Studies* 44:3 (2016), 444.
3   The Fragile States Index is available at https://fragilestatesindex.org/; and the latest Corruption Perceptions Index at the time of writing is at https://www.transparency.org/cpi2018.
4   See, for example, Robert I. Rotberg, (ed.), *When States Fail: Causes and Consequences* (Princeton NJ: Princeton University Press, 2010); Sebastian Mallaby, 'The Reluctant Imperialist: Terrorism, Failed States, and the Case for American Empire', *Foreign Affairs* 81:2 (2002), 2-7; Derick W. Brinkerhoff, 'State fragility and failure as wicked problems: beyond naming and taming', *Third World Quarterly* 35:2 (2014), 333–44.

The second approach views external interventions as the main explanatory factor *for* the phenomenon of state failure, either as a means to control the state's political regime or to serve the strategic interests of the intervening parties.[5] The problem is that the first approach underplays any significant negative role or unintended consequence of foreign intervention in the process of state failure, while the second ignores any ethical and legal motives behind intervention and does not offer an analytical framework to explain how intervention might produce state failure. Thus, it is important to study the mechanisms through which foreign intervention can lead to state fragility and failure. Accordingly, this chapter builds on the conceptual framework provided by Sang Ki Kim relating to third-party intervention in civil wars[6] to analyse the impact of intervention motives, patterns, forms and instruments on strengthening or decreasing the capacity of the state to carry out its functions at the political, security, and economic levels. Figure 7.1 below outlines this conceptual framework.

The theoretical framework that explains the mechanism through which humanitarian intervention could produce state fragility and failure is based on three pillars and is first outlined and then applied systematically to the selected case study of Libya:

- *The determinants of intervention* guide the methods of intervention and the timing of intervention. They include both internal and external determinants. The internal determinants refer to the internal environment that allows intervention, or at least is unable to prevent it. It could also include internal actors who might invite intervention to enhance their relative position within an existing internal conflict. The external determinants of intervention include the external actor/s wishing to intervene in a specific state, either for value-oriented determinants or self-interested material determinants, or some combination of the two. These external actors might intervene through creating an environment within the state that would make it more permissive for intervention, or

---

5   See, for example, Noam Chomsky, *Failed States: The Abuse of Power and the Assault on Democracy* (American Empire Project: Owl Books, 2007); Charles Call, 'The Fallacy of the "Failed State"', *Third World Quarterly* 29:8 (2008), 1491-1507; Sonja Grimm, et al (eds), *The Political Invention of Fragile States: The Power of Ideas* (Abingdon: Routledge, 2014); Mary Manjikian. 'Diagnosis, Intervention, and Cure: The Illness Narrative in the Discourse of the Failed State', *Alternatives: Global, Local, Political* 33:3 (2008), 335–57.

6   See Sang Ki Kim, 'Third-party intervention in civil wars: motivation, war outcomes, and post-war development', (Iowa City IA: Graduate College of The University of Iowa, 2012).

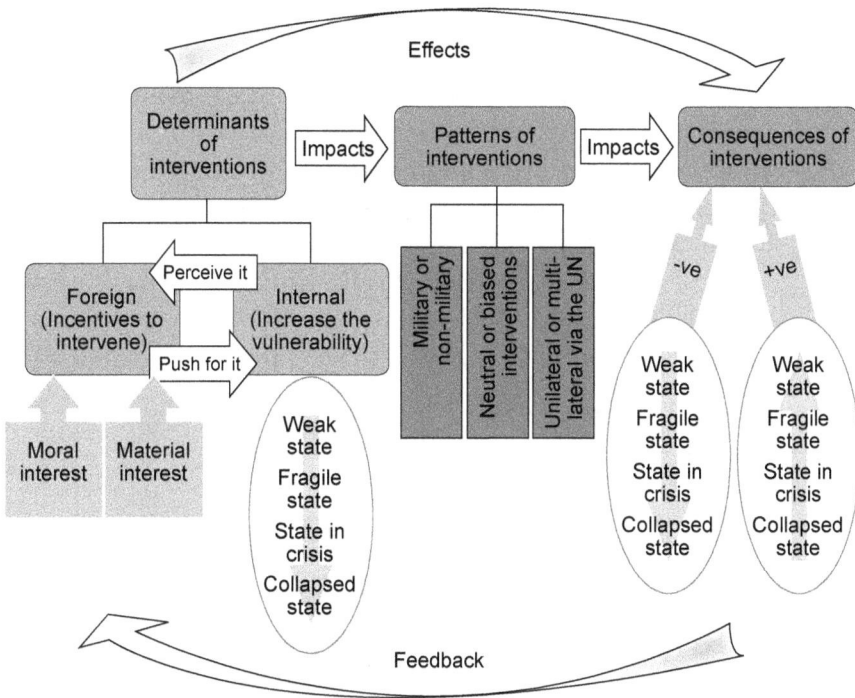

Figure 7.1: Interventions and State Fragility: A Conceptual Framework

they might intervene if they perceive the existing environment represents either an opportunity to pursue interests or a source of risk or threat that needs to be averted.

- *Intervention methods* refers to the mechanisms through which an intervention is conducted. They include the patterns of intervention (biased or neutral intervention), the form of intervention (unilateral intervention or multilateral intervention, authorised by the UNSC or carried out by another international actor), and the instruments of intervention (military or non-military). Intervention methods can influence both the duration and outcome of internal conflict and both positively and negatively affect the state's future capacity to carry out its essential functions in key security, economic, social, political and humanitarian areas.

- *Intervention outcomes* (particularly military victory or negotiated settlement) refers to the impact of the intervention on the affected state's capacity to carry out its mandated functions in the post-conflict period at a variety of aforementioned levels. The intervention might strengthen or weaken a state's capacity to carry out those functions. Table 7.1 describes the state failure typology presented by Paul D. Miller, which is based on combining five types of state failure – anarchic, illegitimate, incapable, unproductive and barbaric – with three ascending degrees of failure: weak, failed/ failing and collapsed. There are many indexes[7] that can capture the general trend of state capacity in performing its functions over time. By comparing these indexes before and after the intervention, changes in state performance can be identified and underscored.

	Security	Legitimacy	Capacity	Prosperity	Humanity
Weak	Unstable	Fragile consensus	Minimal functioning	Poor	Repressive
Failing/ Failed	Violent	Widespread disenfran- chisement	Partial functioning (some collapse)	Destitute	Totalitarian
Collapsed	Anarchic	Illegitimate	Incapable	Unproductive	Barbaric

Table 7.1: Paul D. Miller's Typology of State Failure
Source: D. Miller, 'Strategies of Statebuilding: Causes of Success and Failure in Armed International Statebuilding Campaigns by Liberal Powers', *APSA 2010 Annual Meeting Paper*, 176.

The Internal Determinants of Intervention in Libya

The internal determinants relate to those factors within Libya that increased the vulnerability of the state to foreign intervention. Historically, arguably, Libya has long been a weak state.[8] Moreover, Libya has been an arena of foreign interventions since the nineteenth century. The interventions by the Ottomans and Italians (1911–22) and later by the British, French and Americans (1943–69) resulted in 'strengthening local non-state political

---

7   These include the Fragile States Index, Worldwide Governance Indicators, Human Development Index, Corruption Perceptions Index, Global Peace Index and Global Terrorism Index.
8   Until its independence in 1951, Libya had never been governed by a single entity and the central state was weak, and outside the coastal towns, territory was controlled by local tribes. See Jean-Louis Romanet Perroux, 'The Deep Roots of Libya's Security Fragmentation', *Middle Eastern Studies* 55:2 (2019), 200-224.

identities and alliances' by 'creating incentives and opportunities for provincial powerbrokers to use international patrons to reinforce their authority and prolong their dominance across generations'.[9] It is worth briefly considering here the historical developments within Libya prior to the 2011 intervention by leading Western powers and NATO as a means to contextualising the nature of the Libyan state that the international community could have inherited.

When Muammar Al-Gaddafi came to power in 1969, he demanded the evacuation of British and American military bases and supported what he called revolutionary movements across the world. As a result of such actions, Libya was eventually designated a state sponsor of terrorism, and subjected to sanctions.[10] Gaddafi's distrust of international norms, including those governing sovereignty and statehood, when considered alongside his attempts to consolidate his grip on power, led him to deliberately undermine formal state institutions in favour of informal patronage networks utilising income from Libya's natural resources in the interests of the regime, rather than the state as a formal entity. This resulted in Gaddafi's effective legacy being the continuation of Libya as a weak state with weak institutions over the four decades of his rule.[11]

Gaddafi kept the army weak to prevent the emergence of potential competitors. He sidelined the army and police in favour of parallel and unofficial structures of revolutionary security committees and revolutionary guards responsible primarily for the regime's survival.[12] In addition, the Gaddafi regime entrenched unequal development on a regional basis, which resulted in an escalation of regional identity as an alternative source of legitimacy – and potential challenge to – a centralised (if weak) state. Gaddafi relied on existing tribal structures politically, notably after 1993, in seeking to stabilise his regime by building informal tribal alliances, which resulted in accusations of nepotism and tribal favouritism.[13] During the

---

9   Lisa Anderson, '"They Defeated Us All": International Interests, Local Politics, and Contested Sovereignty in Libya', *Middle East Journal* 71:2 (2017), 239.

10   Anderson, 239.

11   See Lisa Watanabe, 'Libya: In the Eye of the Storm', *CSS Analysis in Security Policy* 193 (2016), 3; Romanet Perroux, 1–25; Edward Randall, 'After Gaddafi: Development and democratization in Libya', *Middle East Journal* 69:2 (2015), 204; Tim Eaton, *Libya's War Economy: Predation, Profiteering and State Weakness* (London: Royal Institute of International Affairs, 2018), 5.

12   See Luis Martínez, 'Libya from Paramilitary Forces to Militias: The Difficulty of Constructing a State Security Apparatus', *Policy Alternatives* (2014), 2; Florence Gaub, *Libya: The Struggle for Security* (Paris: European Union Institute for Security Studies, 2013), 2.

13   Mohamed El-Katiri, *State-building Challenges in a Post-revolution Libya* (Carlisle PA: US Army War College Strategic Studies Institute, 2012), ix.

2000s, there was an attempt at internal reform led by Gaddafi's son, Saif Al-Islam, who was seen by some Western policy makers as the best prospect of effecting political change in Libya.[14] As a result of his increasing influence, the regime began to take a more conciliatory approach to the West after 2003, perhaps aided by existential fears about its own survival following the coerced collapse of the Saddam Hussain regime in Iraq.[15] Consequently, the Gaddafi regime acknowledged responsibility for the 1988 Lockerbie airline bombing and agreed to pay compensation to victims, in addition to further concessions to Western concerns about the level of private control within the Libyan economy. Both the UN and then the US lifted sanctions shortly thereafter, as a move to normalise relations with the regime.

Significantly, the economic reform policy headed by Shukri Ghanem as prime minister in the period 2003-06 resulted in the emergence of a new entrepreneurial class of capitalist reformers within Libyan society. This new class, in addition to a mixed group of businessmen and recently returned Libyan émigrés, saw Saif as a bridge between the old repressive regime and a newer, freer one.[16] Moreover, Saif adopted what could be considered an appeasement policy towards Libya"s main Islamist groups, the Libyan Muslim Brotherhood (LMB) and the Libyan Islamic Fighting Group (LIFG), in the mid–2000s, seeking to neutralise their opposition to the regime. From 2009 to 2010, he spearheaded the release of 946 high-profile Islamist prisoners from Abu Selim prison in return for Islamist recognition of the regime and the renunciation of violence.[17]

In doing so, he effectively championed reform measures that would sow the seeds of what became the 2011 rebellion, as part of the wider so-called 'Arab Spring'. Concerned that they would ultimately be sidelined by the regime's old guard, the new class of capitalist reformers, businessmen and returned Libyan émigrés would join the political front of the rebellion. Furthermore, the released Islamists also assumed leadership or fighting roles in the rebellion against the regime from February 2011 onwards.[18] Ghanem's limited economic reform of cutting state subsidies on basic goods angered many Libyans, giving a banner cause for potential

14 House of Commons Foreign Affairs Committee, *Libya: Examination of Intervention and collapse and the UK's future policy options* (London: The Stationery Office, 2016), 18; Alison Pargeter. *Libya: The Rise and Fall of Gaddafi* (New Haven CT: Yale University Press, 2012), 203.
15 Randall, 207.
16 Ethan Chorin, *Exit the Colonel: The Hidden History of the Libyan Revolution* (New York: Public Affairs, 2012), 94.
17 Chorin, 150.
18 Ibid.

revolt.[19] Moreover, during the 2000s, new lines of corruption emerged that were perceived as benefiting Gaddafi's family members directly,[20] with Wikileaks' release of cables from the US embassy in Tripoli outlining the scale and nature of the familial corruption only increasing the perceived need for regime change, brought about by internal revolt.[21] Finally, as a longer term internal trend, this period of increased opposition, initially even encouraged by a weaker and less popular regime, coincided with the introduction of the internet and social media into Libya, which later played a significant role in the 2011 uprising.[22]

The External Determinants of Intervention in Libya

As noted earlier, external determinants of intervention can include both value-oriented moral determinants and self-interested material ones that represent incentives for foreign actors to intervene. The US played an indispensable political and military role in the 2011 intervention in Libya. However, this was downplayed by the then Obama administration and characterised as 'leading from behind'.[23] France and the UK assumed the lead in pushing for and carrying out military intervention in Libya. However, Paris was possibly too proactive, initiating action one step ahead of its allies, while London was more cautious and found itself repeatedly one step behind the curve.[24]

The UK and US preferred NATO to take command of the military mission after the initial phase, which was primarily US-led, while France,

---

19 Pargeter, 195.

20 Ibid, 207; El-Katiri, 6

21 Ronald Bruce St John, *Libya: From Colony to Revolution* (London: Oneworld Publications, 2017), 83.

22 Masudul Biswas, and Carrie Sipes, 'Social Media in Syria's uprising and post-revolution Libya: an analysis of activists' and Blogger's online engagement', *Arab Media & Society* 19 (2014), 1.

23 Jeffrey Goldberg, 'The Obama Doctrine', *The Atlantic*, April 2016, available at https://www.theatlantic.com/magazine/archive/2016/04/the-obama-doctrine/471525/.

24 France took the lead in launching air strikes, recognising the Libyan National Transitional Council, initiating the International Contact Group on Libya, and most importantly refusing the British suggestion of a military pause in air operations after securing Benghazi. See Madelene Lindström, and Kristina Zetterlund, 'Setting the stage for the military intervention in Libya', *Decisions made and their implications for the EU and NATO* (Stockholm: Swedish Ministry of Defence, 2012), 13; House of Commons Foreign Affairs Committee, available at http://data.parliament.uk/writtenevidence/committeeevidence.svc/evidencedocument/foreign-affairs-committee/libya-examination-of-intervention-and-collapse-and-the-uks-future-policy-options/oral/27184.html.

reviving historical scepticism of the US-dominated organisation even at a time when the Obama administration was seeking to practice what it had preached regarding burden-sharing, including leadership of military operations, preferred a bilateral military command with the UK.[25] While NATO finally led the military operation, France and the UK initiated the International Contact Group on Libya (ICG) as a political coordination mechanism for NATO military operations. Thus, 'the ICG was an international umbrella organisation for those external actors involved in bringing about regime change'.[26] It was significant when considering both the balance between morality and self-interest in what was publicly proclaimed to be a humanitarian action in response to a specific internal crisis, and the end outcomes desired formally and informally by such actors.

The declared rationale of the multilateral military intervention in Libya as expressed publicly was a moral humanitarian purpose under the Responsibility to Protect (R2P) principle. However, it is argued here that the political end goal was regime change. This was pitched as a desirable *indirect* consequence of the intervention, rather than an explicit military end goal, attached to a explicitly humanitarian cause.[27] This political end goal was explicit in French and American official statements, most notably then-French President Nicolas Sarkozy's early position that 'Gaddafi has to go',[28] a position echoed by President Barack Obama.[29] Then UK Prime Minister, David Cameron, was more cautious in articulating a direct political desire to see Gaddafi ousted. General Sir David Richards, then UK Chief of Defence Staff (CDS), admitted later that, after securing Benghazi, 'the humanitarian rationale morphed into a change of regime one'.[30] However,

---

25 Sergei Boeke and J. de Roy van Zuijdewijn, 'Transitioning from military interventions to long-term counter-terrorism policy: The case of Libya (2011–2016)', *Security and Global Affairs* (2016), 26.

26 Saskia M Van Genugten, *Italian and British relations with Libya: pride and privileges 1911–2011* (Baltimore MD: Johns Hopkins University Press, 2012), 348.

27 In addressing this question, Paul Tang Abomo concluded that 'The political goal was to see Gaddafi step down and the military mission evolved to match the political goal. With NATO taking over, the military mission changed from protecting civilians to degrading the military capacities of Libyan forces so that the Libyan people could take charge of their destiny. It practically became a call for regime change'. Paul Tang Abomo. *R2P and the US Intervention in Libya* (Basingstoke: Palgrave Macmillan, 2019), 210.

28 Valentina Pop, 'France says Gaddafi must go, as EU finalises sanctions', *EU Observer*, 25 February 2011, available at https://euobserver.com/news/31876.

29 'Remarks by President Obama and President Sebastian Pinera of Chile at Joint Press Conference, March 21, 2011', available at https://obamawhitehouse.archives.gov/the-press-office/2011/03/21/remarks-president-obama-and-president-sebastian-pinera-chile-join-press-.

30 House of Commons Foreign Affairs Committee, oral evidence.

even with this more cautious evolutionary explanation, it was clear that the political end goal was regime change in Libya, and this was confirmed in a joint op-ed by the three heads of government published on 15 April 2011. In it, they argued that 'our duty and mandate under UNSC Resolution 1973 is to protect civilians. It is not to remove Gaddafi by force. However, it is impossible to imagine a future for Libya with Gaddafi in power'.[31]

It is worth briefly considering the stated and implicit motivations of the leading members of the Western intervention. In the case of the US, it perceived the intervention in Libya as practicable, based on the condition that it would be (self-) limited in time and space. The US saw no direct vital interest to intervene in Libya, beyond a moral obligation to prevent an assumed looming genocide – in and of itself a disputed assessment in academic circles[32] – as well as to 'pay back' its allies, France and the UK, for their support in Afghanistan. In addition, the US may have had an economic interest in the Libyan oil industry – it was the first country to buy oil from the Libyan National Transitional Council (NTC) after Gaddafi's overthrow. President Sarkozy's plans for the intervention in Libya were driven by multiple motives, according to an official US source. These included a 'desire to gain a greater share of Libya oil production,[33] increase French influence in North Africa, improve his internal political situation in France, provide the French military with an opportunity to reassert its position in the world, [and] address the concern of his advisors over Gaddafi's long term plans to supplant France as the dominant power in Francophone Africa'.[34] In the case of the UK, the intervention met three stated conditions: demonstrable need, legal basis and regional support. Additionally, David Cameron – in line with a longer heritage of UK leaders, notably Tony Blair

---

31 Joint op-ed by President Obama, Prime Minister Cameron and President Sarkozy, 'Libya's Pathway to Peace', White House Press Release, 2011, available at https:// obamawhitehouse.archives.gov/the-press-office/2011/04/14/joint-op-ed-president-obama-prime-minister-cameron-and-president-sarkozy.

32 See Alan J. Kuperman, 'NATO's Intervention in Libya: A Humanitarian Success?', in Aidan Hehir and Robert Murray (eds), *Libya, the Responsibility to Protect and the Future of Humanitarian Intervention* (Basingstoke: Palgrave Macmillan, 2013).

33 Seven days after the passing of UN Resolution 1973, France is said to have concluded a deal with the NTC stipulating that, together with Qatar, it would be guaranteed 35 per cent of oil contracts in exchange for its total and permanent support of the Transitional Council. See Vittorio De Filippis, 'Pétrole: l'accord secret entre le CNT et la France', *Liberation*, 1 September 2011, available at https://www.liberation.fr/planete/2011/09/01/petrole-l-accord-secret-entre-le-cnt-et-la-france_758320.

34 US Department of State, 'H: France's client and Q's gold. Sid' [sic], 2 April 2011, available at https://www.foia.state.gov/Search/results.aspx?searchText=C05779612&beginDate=&end Date=&publishedBeginDate=&publishedEndDate=&caseNumber=.

– saw Libya as an opportunity to demonstrate his leadership on global security issues, remain relevant to the US when it called for someone else to take the lead in Libya – as part of the so-called 'special relationship', even if they would be operating primarily instead of rather than alongside – and meet wider economic interests regarding future Libyan oil and gas production.[35] In essence, while circumstances differed, the underlying mix of economic interest, status, statesmanship, alliance formulation and leadership, coupled with both immediate moral and wider, longer term strategic concerns is a recognisable mix in all three cases when considering the impetus for intervention.

In fact, the timing of the intervention is very revealing, coming barely one month after the rebellion against Gaddafi broke out. The stated US initial reluctance to intervene changed after intelligence assessments showed clear evidence that Gaddafi's forces would prevail in the end.[36] Credibility, as is so often the case, became a core factor in decision making. Coalition operations began merely two days after the passing of United Nations Security Council Resolution (UNSCR) 1973, attacking Gaddafi's forces about 40 miles from their intended target of Benghazi.[37] Had Gaddafi's forces not been stopped, the Libyan regime would likely have been able to consolidate its position in the east of the country, while the rebels would have been seriously undermined, because Benghazi was the base of what would become the NTC.[38]

The Mechanism of Intervention in Libya

Intervention methods reflect the interveners' motives and goals and the impact such approaches have on the intervention's outcome.[39] Key in this case is how the conduct of the intervention conformed with its declared and mandated ethical goal, under the rubric of the R2P principle – itself

---

35 See 'France, U.K. Have Differing Motives for Intervening in Libya', *Forbes* magazine, 29 March 2011, available at https://www.forbes.com/sites/energysource/2011/03/29/france-u-k-have-differing-motives-for-intervening-in-libya/#7dc0ee05ad53; also Lindström, and Zetterlund, 38.
36 Abomo. 243.
37 House of Commons Foreign Affairs Committee, 17.
38 Christina Goulter, 'The War in Libya: The Political Rationale for France' in D. Henriksen and A. K. Larssen (eds), *Political Rationale and International Consequences of the War in Libya* (Oxford: Oxford University Press, 2016), 54.
39 Sang Ki Kim, 5.

drawn from wider 'just war' principles and authorised by UNSCR 1973.[40] The most notable principles to be considered here are: just cause, right intention, last resort and proportional means.

Just Cause

In the Libyan case, while there might be a just cause for military intervention to prevent attacks against civilians, the position is not as clear-cut as some Western narratives suggest. Reinforcing the idea of a 'just cause', the first report of the UN Human Rights Council (HRC) Commission of Inquiry on Libya in June 2011 noted that government forces there had committed serious violations 'amounting to war crimes' and 'crimes against humanity'.[41] However, in its second report in March 2012, it concluded that the *thuwar* (anti-Gaddafi forces) had also committed crimes of a similar scale.[42] This was corroborated in an Amnesty International investigation in June 2011, which claimed that Western media coverage had ignored evidence the protest movement exhibited a violent aspect from its initial stages onwards.[43] Moreover, many Western policy makers 'proposed that Gaddafi would have ordered the massacre of civilians in Benghazi if those forces had been able to enter the city',[44] but this proposition, according to the findings of the UK House of Commons Foreign Affairs Select Committee, 'was not supported by the available evidence'.[45] It concluded that UK Government strategy in Libya was founded on 'erroneous assumptions and an incomplete understanding of the evidence', due – ironically given the political difficulties that engulfed a previous government with regard

---

40 Elizabeth O'Shea, 'Responsibility to Protect (R2P) in Libya: ghosts of the past haunting the future', *International Human Rights Law Review* 1:1 (2012), 178.

41 UN General Assembly, *Human Rights Council: Report of the International Commission of Inquiry to investigate all alleged violations of international human rights law in the Libyan Arab Jamahiriya* (UN Doc. A/HRC/17/44), 2011, 7, available at http://www2.ohchr.org/english/bodies/hrcouncil/docs/17session/A.HRC.17.44_AUV.pdf.

42 UN General Assembly, *Human Rights Council: Report of the International Commission of Inquiry to investigate all alleged violations of international human rights law in Libya* (UN Doc. A/HRC/19/68), 2012, 196–7, available at
http://www.ohchr.org/Documents/HRBodies/HRCouncil/RegularSession/Session19/A.HRC.19.68.pdf.

43 International Crisis Group, 'Popular Protest in North Africa and the Middle East (V): Making Sense of Libya', *Middle East/North Africa Report* 107 2011, 4, available at https://d2071andvip0wj.cloudfront.net/107-popular-protest-in-north-africa-and-the-middle-east-v-making-sense-of-libya.pdf.

44 House of Commons Foreign Affairs Committee, 15.

45 Ibid., 14.

to Iraq – to incomplete intelligence.[46] This conclusion was also highlighted by some officials within the US intelligence community, who noted that 'there was no specific evidence of an impending genocide in Libya in spring 2011', believing that advocacy for intervention against the Libyan regime rested more on 'speculative arguments of what might happen to civilians than on facts reported from the ground'.[47]

Right Intention

The right intention for intervention under the declared R2P norm was to protect civilians, as was demonstrated in UNSCR 1973, which provided the legal basis 'authorizing Member States to take all necessary measures to protect civilians', while 'excluding a foreign occupation force of any form on any part of Libyan territory'[48] and confirming NATO's political end goals for its Libya campaign. However, and unsurprisingly given the underlying structures and strictures of international law, regime change was *not* part of the UNSC mandate, implicitly or explicitly.[49] It is possible that NATO's intervention was born of a desire to protect civilians. However, within a few weeks of launching the operation, NATO took actions that were unnecessary or inconsistent with protecting civilians, but which fostered regime change.[50] For instance, 'the combination of coalition airpower with the supply of arms, intelligence and personnel to the rebels guaranteed the military defeat of the Gaddafi regime'.[51] Furthermore, in early 2011, foreign governments backing the NTC only had a vague idea who they were providing with weapons, intelligence and other support. Some policy makers raised concerns about

---

46 Ibid., 15. The committee's conclusions underscored that Gaddafi forces had retaken towns from rebels without attacking civilians in early February 2011.

47 See 'Hillary Clinton's 'WMD moment': US intelligence saw false narrative in Libya', *Washington Times*, 29 January 2015, available at: https://www.washingtontimes.com/news/2015/jan/29/hillary-clinton-libya-war-genocide-narrative-rejec/.

48 United Nations, Security Council, Resolution 1973, 17 March 2011, available at https://undocs.org/S/RES/1973(2011).

49 Heinz Gärtner, 'The Responsibility to Protect (R2P) and Libya', *Australian Institute for International Affairs* 1090:7 (2011), 111.

50 Less than two weeks into the intervention, NATO began attacking Libyan forces that were retreating and thus not a threat to civilians. Rather than pursuing a ceasefire, NATO and its allies aided the rebels who rejected that peaceful path and sought instead to overthrow Gaddafi: Kuperman, 197.

51 According to the House of Commons Foreign Affairs Committee, 'If the primary object of the coalition intervention was the urgent need to protect civilians in Benghazi, then this objective was achieved in less than 24 hours', House of Commons Foreign Affairs Committee, 17.

the fact those fighting on the side of the opposition seemed to include extremist elements, particularly of an Islamist nature.[52]

Moreover, NATO's support for the rebels seeking regime change was premised on Gaddafi being identified as a *permanent* threat to the human rights of Libyan citizens. Given this, the fact that rebel forces were responsible for widespread human rights abuses raises difficult questions for those who supported the intervention. Furthermore, it seems that there was an exaggeration of the initial death toll that provoked the intervention, with the putative threat of further massacres as important when justifying action as the actual numbers already killed.[53] In contrast to the NATO Secretary–General's claim that NATO-led forces had prevented a massacre and saved countless lives,[54] Alan Kuperman argues that, as a result of NATO's intervention, Libya's war lasted 36 weeks instead of six and it magnified the death toll in Libya by between seven and 27 times.[55] He further argues that the expectation of such intervention helped trigger or escalate the Libyan rebellion that then provoked government retaliation and thereby endangered civilians in a wider conflict. Such a potential dynamic is known as the 'moral hazard' of humanitarian intervention.[56] This point was also underscored by Rajan Menon, who argues that the anti-Gaddafi fighters turned to violence at an early stage, lobbied actively for external intervention following a crackdown by the regime and publicised inflated figures on civilian deaths.[57]

Last Resort

The last resort principle in the Libyan case is an even more debatable one. Some scholars argue that military action in Libya was preceded by a range of non-military measures that sought to persuade the Gaddafi regime

---

52 The rebels constituted an unorganised and rather undisciplined group of strange bedfellows united around one purpose only: killing Gaddafi and bringing the downfall of his regime. This brought together former monarchists, liberals, federalists, Islamists and Jihadists. See Van Genugten, 152.

53 Post-war surveys by the Red Cross and other humanitarian agencies found a total of fewer than 3,500 accountable in major cities, while the NTC claimed 30-50,000 deaths. See Michael W Doyle, 'The politics of global humanitarianism: The responsibility to protect before and after Libya', *International Politics* 53:1 (2016), 23.

54 Simon Adams, *Libya and the Responsibility to Protect* (New York: Global Centre for the Responsibility to Protect, 2012), 15.

55 Kuperman, 206.

56 Ibid, 208-9.

57 Rajan Menon, *The Conceit of Humanitarian Intervention* (Oxford: Oxford University Press, 2016), 14.

to stop the killing. All the steps considered in UNSCR 1970, including the freezing of assets, an arms embargo, travel bans and referrals to the International Criminal Court (ICC), while coercive, were peaceful.[58] It was only when these measures failed that the use of military force was duly considered.[59] On the other hand, others contend that the non-military options for the peaceful resolution of the crisis were not given the required amount of time and the military option was prioritised at the expense of a possible comprehensive political solution to the Libyan crisis.[60] The African Union (AU) had been able to reach an outline agreement with Gaddafi.[61] However, the NATO members on the UNSC had actively 'blocked' the AU's attempts to peacefully resolve the Libyan conflict by refusing to allow an AU delegation to fly to Libya to 'begin the process of mediating a peaceful resolution'.[62] Moreover, the ICC referral may be one of the reasons Gaddafi refused to take safe haven in another state.[63] This is an important point, as Gaddafi might have been seeking an exit from Libya in February and March 2011.[64] Had the conflict ended earlier with a negotiated settlement leading to Gaddafi's exit, this could have saved lives and also led to a more stable post-conflict Libya.

Proportional Means

The judgement over the use of proportional means in the Libyan case is also contested. States such as Brazil, India, China, Russia and South Africa have argued that NATO's actions exceeded the minimum intervention necessary to secure the stated human protection objective. As evidence, they point to the perceived violation of the terms of the arms embargo outlined in UNSCR 1970 by some Western states, who supplied arms to anti-Gaddafi forces and turned a blind eye to the supply of weapons to

58 United Nations, Security Council, Resolution 1970, 26 February 2011, available at https://undocs.org/S/RES/1970(2011).

59 Adams, 12.

60 Christopher Zambakari, 'The misguided and mismanaged intervention in Libya: Consequences for peace', *African Security Review* 25: 1 (2016), 48.

61 Marwan Hameed, 'Responsibility to Protect: The Use and The Abuse' (New York: CUNY Academic Works, 2014), 44, available at https://academicworks.cuny.edu/cc_etds_theses/544/.

62 Adams 2012, 15; Thabo Mbeki, 'We should learn from Libya's experiences', *Mail & Guardian Online*, 5 November 2011, available at: https://mg.co.za/article/2011-11-05-mbeki-we-should-learn-from-libyas-experiences.

63 Christo Odeyemi, 'R2P intervention, BRICS countries, and the no-fly zone measure in Libya', *Cogent Social Sciences* 2:1 (2016), 9.

64 House of Commons Foreign Affairs Committee, 19.

the rebels by other external actors engaged in the conflict, notably Qatar. In June 2011, France admitted to supplying weapons to the rebels and Qatar allegedly admitted later that it had hundreds of troops fighting on the ground against Gaddafi's forces.[65] Other forms of support from France and the UK included 'providing battleground leadership advice during the final rebel offensive on Tripoli and Sirte',[66] sending in military advisers, Special Operations Forces and para-military intelligence officers to help 'train and arm the rebels'[67] as well as 'deploying attack helicopters to the conflict'.[68] This was a direct violation of UNSCR 1970 and was not in keeping with the spirit of the civilian protection mandate represented in Resolution 1973. They also questioned the implementation of the no-fly zone that had been authorised by UNSCR 1973, suggesting that NATO's military actions visibly gave preferential support to the rebels by no longer acting solely as a defensive shield for populations at risk, but as the NTC's 'air force'.[69] Moreover, the tactical use of NATO airpower to support the rebel offensive against Tripoli, the bombing of Libyan TV and the attempted assassination by drone of Gaddafi himself arguably strained the logic of R2P.[70] Furthermore, on 20 October 2011, it was a US drone and French aircraft that attacked a convoy of regime loyalists trying to flee Gaddafi's hometown of Sirte. Gaddafi was injured in the attack, captured alive and then extra-judicially executed by rebel forces.[71]

Overall, the intervention was multilateral, firstly through a US-led international coalition that was transferred later to NATO leadership, but which, significantly, both for alliance solidarity and for strictly applying the spirit and letter of the law, included some *unilateral* actions outside the NATO

---

65 Adams, 12.

66 Ibid, 12.

67 See Eric Schmitt and Steven Lee Myers, 'Surveillance and Coordination with NATO aided Rebels', *New York Times*, 22 August 2011, available at http://www.nytimes.com/2011/08/22/world/africa/22nato.html; Ian Traynor and Richard Norton-Taylor, 'Libya's 'mission creep' claims as UK sends in military advisors', *The Guardian*, 19 April 2011, available at https://www.theguardian.com/world/2011/apr/19/libya-mission-creep-uk-advisers; Marcus Mohlin, 'Cloak and Dagger in Libya: The Libyan Thuwar and the role of allied special forces', in Engelbrekt, Kjell, Marcus Mohlin, and Charlotte Wagnsson (eds), *The NATO Intervention in Libya: Lessons Learned from the Campaign* (Abingdon: Routledge, 2013), 195.

68 Michael W. Kometer and Stephen E. Wright, 'Winning in Libya: By Design Or Default?', (Paris: Institut Français des Relations Internationales), 2013, 26.

69 Adams, 12.

70 Doyle, 24.

71 Micah Zenko 'The Big Lie About the Libyan War', Foreign Policy, 22 March 2016, available at https://foreignpolicy.com/2016/03/22/libya-and-the-myth-of-humanitarian-intervention/.

chain of command. The pattern of intervention was a biased intervention against Gaddafi forces using a combination of coalition airpower with the supply of arms, intelligence and personnel to the rebels, which shifted the military balance in the Libyan civil war in favour of the perceived 'good guys'. The instruments of intervention were military and also non-military by imposing sanctions and an arms embargo on the Gaddafi regime. Such an intervention, whether by design or default, proved both decisive and divisive, ensuring the victory of Gaddafi's opponents and a more chaotic process of extended 'regime change', but the NATO intervention was ended without any stabilisation phase in place, as there was no expressed desire, internally or externally, for long-term involvement in costly state-building efforts,[72] ultimately leaving the fate of the state to the unknown and more extremist elements. In effect, Western leaders hoped the state would stabilise itself, without acknowledging the nature and duration of their intervention had made such a prospect highly problematic, if not wholly unlikely.

The Outcomes of Intervention

In the aftermath of the intervention, Libya turned from being a weak state under the Gaddafi regime into arguably a failed state, characterised by a violent power struggle for control of what remained of the security sector and key economic assets, which led to the establishment of parallel security and economic institutions. The vacuum caused by ousting the Gaddafi regime left the fate of the Libyan state in the hands of the rebels, who were composed mainly of two opposing camps. On one side, there was the reformist camp, which included members of the post-2003 reform wing of the Gaddafi government, and, opposing them, a conservative camp, which included elements of the Islamist forces of the LMB and LIFG. It is worth considering how such splits have affected the nature of Libyan governance in the post-intervention period, to determine the wider implications of the manner and duration of Western direct involvement.

The division had political and military implications. The two competing camps – prominent within the intervention to topple Gaddafi – reappeared during the first national elections on 7 July 2012 to elect

---

72 Kometer and Wright, 30.

the General National Congress (GNC). The civilian-oriented camp were represented by the coalition of National Forces Alliance (NFA), while the more conservative, Islamic-oriented camp were represented by the Justice and Construction Party (for the LMB) and the Nation Party (for the LIFG). The relative majority obtained by the NFA in 2012 elections demonstrated at least an initial rejection of the Islamist parties. However, the NFA was not able to rule due to coercion and intimidation by its Islamist opponents. Using violence and political manoeuvres, including the adoption of a controversial political isolation law by the GNC on 5 May 2013 at gunpoint, the Islamists were still able to dominate the GNC and control the government. A former US special envoy for Libya lamented that 'Western diplomats working on Libya generally agreed that their biggest collective mistake was the failure to take action in May 2013 to recognize the lustration law for what it was: a power grab'.[73] Not only did this highlight the singular lack of a monopoly over the legitimate use of force, a necessary condition for effective statehood, but it also undermined the nascent credibility of Libyan democracy, which required decisions to be taken by non-violent means.

As a response to increasing popular discontent, the GNC agreed to hold new parliamentary elections in June 2014. Despite Islamist parties suffering a devastating loss at the ballot box, they sought to retain control by force, taking control of the capital militarily and having the Islamist-dominated GNC refuse to hand over power to the House of Representatives (HOR), the newly elected parliament. In effect, with the abandonment of even the pretence at democratic civilian discourse, two new rival military coalitions were formed after May 2014. 'Operation Dignity', centred in eastern Libya and aligned with the HOR, was led by the Libyan National Army (LNA) under the command of General Khalifa Haftar. Their declared aim was the defeat of Islamists in the cities of eastern Libya. On the other side, a coalition of Islamist militias from western Libya – dominated by those from the city of Misrata – created the 'Libya Dawn' coalition aligned with the former GNC and its National Salvation Government. In addition to these rivalries, Libya's internal split in 2014 created an arena for regional rivalries, with Egypt and the United Arab Emirates (UAE) backing the Tobruk-based HOR and Qatar, Sudan and Turkey standing behind their Islamist, Tripoli-based rival.

---

73 Jonathan M. Winer, 'Origins of the Libyan Conflict and Options for Its Resolution', *Middle East Institute Policy Paper 2019-4*, February 2019, available at https://www.mei.edu/sites/default/files/2019-03/Origins_of_the_Libyan_Conflict_and_Options_for_its_Resolution.PDF.

While the international community at large has not been passive in this area, it is clear that their ability to effect positive change on the ground is limited by a continued lack of political will, knowledge and credibility. In January 2015, the UN launched the negotiations that would ultimately produce a Libyan Political Agreement (LPA) by the end of that year. It created the Government of National Accord's (GNA) Presidential Council (PC), which took office in Tripoli in March 2016 and was tasked to form a government of national accord and an advisory High State Council of former GNC members. The HOR was supposed to act as the sole legislative authority in the country and approve a unity government. While the UN-sponsored LPA's aim was to create a power-sharing deal to overcome institutional and military fractures, in reality they made an already difficult situation more complex, by seeking to introduce another rival claimant to legitimacy and governance. The HOR authority saw the UN and the talks' Western backers as biased toward the Islamist-dominated GNC, believing the accord would produce another Tripoli-based government dominated by Islamist militias, which would use the implementation of its security arrangements to sideline Haftar. Regardless of this internal opposition, the agreement was pushed by the US, France and the UK, raising further concerns regarding the LPA. By January 2016, they had recognised the PC as Libya's executive and had stopped engaging with the HOR Government. Moreover, certain states – such as the US, the UK and Italy – argued the PC should receive external military aid.[74] As a consequence, the HOR refused to endorse the LPA, PC and proposed GNA government in a no-confidence motion passed on 22 August 2016. In contrast, the Tripoli-based Islamist camp regarded the UN political process as a channel through which they could remain politically relevant,[75] particularly as the new government was to be seated in Tripoli, which is under the control of their associated militia and armed groups.

As a consequence, and repeating a longer historical trend, the PC's control of the capital and of key ministries was limited from the outset.[76] The Tripoli-based Islamist militias effectively continue to control the government.[77] The PC was not able to convene and six of its nine members

74 International Crisis Group, 'The Libyan Political Agreement: Time for a Reset', *Middle East and North Africa Report* 170 2016, 23, available at https://d2071andvip0wj.cloudfront.net/170-the-libyan-political-agreement.pdf.

75 Lisa Watanabe, *Islamist Actors: Libya and Tunisia* (Zurich: ETH, 2018), 18.

76 Eaton, 5.

77 Wolfram Lacher, and Alaa Al-Idrissi, 'Capital of Militias', Small Arms Survey *Briefing Paper* (2018), 16, available at http://www.smallarmssurvey.org/fileadmin/docs/T-Briefing-Papers/SAS-SANA-BP-Tripoli-armed-groups.pdf.

became inactive over time. The LPA expired on 17 December 2017 and, while it was extended later via UNSC resolutions, it was clear the GNA and the LPA were unable to establish a new governance structure that would unify state institutions. In the same period, the new UN envoy to Libya, Ghassan Salamé, launched the UN Action Plan (UNAP) for Libya on 20 September 2017, aiming optimistically to create the necessary conditions for the completion of Libya's post-conflict transition by restructuring the current government, convening an inclusive national conference and preparing for elections before September 2018. This plan was, at best, an international community 'face-saving' plan for Libya and already faced serious obstacles. The three tracks for amending the LPA – convening a Grand National Conference, constitution-building and general elections – are either deadlocked or nearing complete collapse due to the continued zero-sum game being played on the ground by Libyan actors. Moreover, Libyan parties have yet to agree on which elections (presidential and/or parliamentary) to hold and in what sequence, while neither the legal nor the constitutional framework is in place at the time of writing.[78]

In May 2018, France, seeking to revamp the UNAP, invited four Libyan leaders, including Sarraj and Haftar, to Paris to sign off on a plan to adopt the necessary electoral laws to hold elections in December.[79] On 8 November 2018, Salamé recalibrated UNAP, which called for a Libyan-led National Conference to be held in the first weeks of 2019 and the electoral process to begin in the spring. In November 2018, Italy had a go at reviving the UNAP in Libya, organising a meeting in Palermo attended by delegations of three different, competing Libyan authorities, where they agreed on adopting the referendum law to complete the constitutional process, holding elections by spring 2019 and, crucially, respecting its results.[80] Unsurprisingly, none of these internationally sanctioned deadlines were met.

While such political debates took place in external actors' centres of power, the reality on the ground in Libya continued to worsen, as the rivalry between the civilian-oriented camp and Islamic-oriented camp ensured that Libya's internationally-approved transitional authorities were neither

---

78 International Crisis Group, 'Making the Best of France's Libya Summit', *Middle East and North Africa Briefing* 58 2018, 4, available at https://d2071andvip0wj.cloudfront.net/b058-making-the-best-of-frances-libya-summit.pdf.
79 'Political statement on Libya. Joint Statement by Fayez al-Sarraj, Aguila Saleh, Khalid Meshri, Khalifa Haftar', Paris, 29 May 2018, available at https://onu.delegfrance.org/Political-statement-on-Libya.
80 'Palermo Conference for and with Libya: Conclusions', 12/13 November 2018, available at http://www.governo.it/sites/governo.it/files/conference_for_libia_conclusions_0.pdf.

capable of exercising any effective form of sovereignty over the territory nor of holding a monopoly over the legitimate use of force, the most basic and fundamental elements of statehood, regardless of cultural and ideological differences and sensitivities. Within the context of this rivalry, competition over security sector institutions is both a means to an end to exert political influence or gain control over economic assets—and an end in itself.[81] Inheriting an already weakened situation – where Gaddafi's parallel security sector had been destroyed or scattered and the regular Libyan armed forces had already partially disintegrated when the regime fell[82] – the two camps' reactions differed as to the way forward, over a long-standing concern over whether to reform or completely rebuild. The civilian-oriented camp considered the former military and what remained of Gaddafi's formal security apparatus to be the backbone on which any new institutions should be rebuilt. In contrast, the Islamist-oriented camp wanted to build wholesale a new army and a new police force, built upon integrating their loyal militia wholesale into newly-formed formal structures.[83] The most notable parallel Islamist-leaning security forces established in this regard were the Supreme Security Committee, which was considered the parallel police force, and the Libya Shield Forces, which was considered the parallel army.[84] Consequently, hybrid security institutions emerged, combining formal and informal elements and allowing competing interests and loyalties to prosper. Although Libya's hybrid units continue to change labels and institutional affiliation, their political power brokers and their interests have been largely constant. These rivalries render the notion of loyalty to the state meaningless for almost all parties. This explains why successive

---

81 The rivalry over controlling the security sector existed inside the Defence and Interior Ministries, for example, with ministers and their deputies representing competing political camps. In turn, these rivalries thwarted attempts to formulate and implement policies and translated into institutional deadlock. See Wolfram Lacher, and Peter Cole, 'Politics by other means: Conflicting interests in Libya's security sector', Small Arms Survey *Briefing Paper* (2014), 25 available at http://www.smallarmssurvey.org/fileadmin/docs/F-Working-papers/SAS-SANA-WP20-Libya-Security-Sector.pdf.

82 Virginie Collombier, 'Make Politics, Not War: Armed Groups and Political Competition in Post-Gaddafi Libya', in Bassma Kodmani and Nayla Moussa (eds), *Out of the Inferno? Rebuilding Security in Iraq, Libya, Syria, and Yemen* (Paris: Arab Reform Initiative, 2016), 57.

83 Collombier, 59.

84 Naji Abou Khalil And Laurence Hargreaves, *Perception of security in Libya and institutional and revolutionary actors* (Washington DC: United States Institute of Peace, 2015), available at https://www.files.ethz.ch/isn/191016/PW108-Perceptions-of-Security-in-Libya.pdf. Moreover, the Islamists established an Integrity and Reform Commission for the armed forces in late June 2013. The Commission aimed to exclude officers who had supported Gaddafi and retiring senior army officers. The Commission's efforts fuelled discontent within the army and contributed to the formation of the faction now led by General Haftar. See Lacher and Cole, 27.

transitional governments have failed to implement proper disarmament, demobilisation and reintegration or security sector reform processes.

Such rivalry does not just manifest itself in the political and security spheres, but also over the crucial area of natural resources, with all the potential linkages between the three areas. Libya is a petro-state in which oil and gas revenues account for around 96 per cent of state revenues.[85] Two-thirds of hydrocarbon production comes from the east. Competition for control of hydrocarbon resources, the infrastructure to exploit them and the revenues derived from their sale are a central driver of the conflict.[86] This competition has manifested itself in a military struggle over the so-called 'oil crescent', a strip along the eastern coast of the Gulf of Sirte, where four of Libya's six hydrocarbon export terminals are located and through which more than 50 per cent of its crude oil exports leave the country. This has led to several oil terminal crises. For instance, in 2016 and 2018, Western countries spearheaded by the US intervened to guarantee the uninterrupted flow of oil from Libyan terminals in the east.[87] They put massive pressures on Haftar, by threatening to subject him to sanctions, in order to allow oil sales from eastern terminals to be managed by the Tripoli-based company under Islamist control.

Furthermore, the international efforts to end Libya's war have ended up reinforcing Tripoli-based institutions' monopoly on these matters, rather than the sharing and managing of oil resources as a basis for a power-sharing settlement. The UNSC backed LPA conferred international recognition upon the PC and recognised the Tripoli-based Central Bank of Libya (CBL) and the Tripoli-based National Oil Company (NOC) as the sole legitimate institutions.[88] However, the Tobruk-based HOR once again refused to recognise the LPA or the resulting government in Tripoli. This has led to several management crises within these key institutions, where the HOR in the aftermath of the elections in 2014 appointed a new governor of the CBL, a new chairman of the Libyan Investment Authority (LIA) and new management of the Libyan Post, Telecommunication and Information Technology Company (LPTIC). However, these appointments were not

---

85 Eaton, 23.

86 International Crisis Group, 'The Prize: Fighting for Libya's Energy Wealth', *Middle East and North Africa Report* 165, 2015, 1-3, available at https://www.files.ethz.ch/isn/195212/165-the-prize-fighting-for-libya-s-energy-wealth.pdf.

87 Winer, 20.

88 International Crisis Group, 'After the Showdown in Libya's Oil Crescent', *Middle East and North Africa Report* 189, 2018, 3, available at https://d2071andvip0wj.cloudfront.net/189-after-the-showdown-in-libyas-oil-crescent.pdf

able to take effect because the international community and the Islamist camp rejected it. In addition, the Tripoli-based CBL refused to disburse most funds directly to the eastern authorities,[89] which led the rival east-based government to establish its own parallel Central Bank (located in Benghazi and al-Beyda), and National Oil Company (based in Benghazi).[90] A HOR-aligned LPTIC headquarters was established in Malta in 2014 and led to legal battles in 2015 between LIA's rival managements in Libya, the UK, Malta and Italy.[91]

Conclusion

This chapter has investigated the relationships between foreign intervention and state fragility and failure via the prism of the Libyan case study. Libya had long been a weak state with weak institutions, which persisted during the Gaddafi regime. Following the Western intervention during 2011, Libya turned into a failed state. Foreign intervention contributed to Libya's state failure regardless of its humanitarian reasoning. As demonstrated, this came about as a result of a range of factors captured in the model discussed earlier, including the motives of interveners, the mechanism of the intervention and its outcome. There was a possibility of negotiated settlement either through the AU proposal of ceasefire or Gaddafi's potential safe exit, but it was not properly or effectively considered. The motives of the interveners in Libya were mainly self-interest, rather than their declared humanitarian purposes, and centred around regime change. The instruments of intervention were both military and non-military (imposing economic sanctions and an arms embargo) under authorisation from the UNSC. Moreover, the pattern of intervention was a biased one that altered the balance of power in favour of the intervener (supporting the rebels) with the aim of helping them achieve a quicker victory rather than encouraging a potentially more stable negotiated settlement. The timing of the intervention was crucial, as it took place only two days after the passing of UNSC Resolution 1973, on the pretext of an overstated threat against civilians.

Having ousted the Gaddafi regime, the military intervention ended with the faint hope that Libya would stabilise itself. In the post-Gaddafi

---

89 International Crisis Group 2015, 19.
90 Ibid., 2018, 3.
91 Eaton, 25.

era, two main rival camps emerged, to compete politically when it suited them and violently when necessary, across the still-limited arena of Libyan statehood. Libya has effectively been split into two regionally based rival governments and parliaments, with sovereignty and the people of Libya the ultimate victims. While less active than in the intervention phase, the international community – in various ways – has sought to stabilise the political situation, although, once again, it is easy to argue that their main motivation has been based on self-interest. Biased in support of the Tripoli-based Islamist camp and aiming to ensure the sole legitimacy of key financial institutions under their control to secure existing oil contracts, the end result has been the creation of a third government (the GNA) that is internationally recognised, but has little if any control, even in the capital. The haste of the international community to hold elections in Libya in the aftermath of the 2011 intervention to confer legitimacy on their new governmental partners, even while acknowledging the severe lack of capacity to govern, was part of the problem. Elections that are essentially unrecognised, and which lack even the basic necessary legal framework and technical arrangements, have further complicated an already complex situation, with competing cries for legitimacy drowned out by the sounds of gunfire. Had the motives and pattern of external intervention been different – with a more genuine and sustained emphasis on protecting civilians and reaching a negotiated settlement – the ultimate outcome of the intervention might have been different. This is a very costly lesson for the Libyan people, and one that has yet to be learned by the international community.

References

Abou Khalil, N. and Hargreaves, L. *Perception of Security in Libya and Institutional and Revolutionary Actors*. Washington DC: United States Institute of Peace, 2015.

Adams, S. *Libya and the Responsibility to Protect*. New York: Global Centre for the Responsibility to Protect, 2012.

Anderson, L. '"They Defeated Us All": International Interests, Local Politics, and Contested Sovereignty in Libya, *Middle East Journal* 71:2, 2017.

Biswas, M. and C. Sipes, C. 'Social Media in Syria's uprising and post-revolution Libya: an analysis of activists' and Blogger's online engagement', *Arab Media & Society* 19, 2014.

Brinkerhoff, D.W. 'State fragility and failure as wicked problems: beyond naming and taming', *Third World Quarterly* 35:2, 2014.

Call, C. 'The Fallacy of the "Failed State"', *Third World Quarterly* 29:8, 2008.

Chomsky, N. *Failed States: The Abuse of Power and the Assault on Democracy*. American Empire Project: Owl Books, 2007.

Chorin, E. *Exit the Colonel: The Hidden History of the Libyan Revolution*. New York: Public Affairs, 2012.

Collombier, V. 'Make Politics, Not War: Armed Groups and Political Competition on Post-Qaddafi Libya', in B. Kodmani and N. Moussa (eds), *Out of the Inferno? Rebuilding Security in Iraq, Libya, Syria, and Yemen*. Paris: Arab Reform Initiative, 2016.

Doyle, M.W. 'The politics of global humanitarianism: The responsibility to protect before and after Libya', *International Politics* 53:1, 2016.

Eaton, T. 'Libya's War Economy: Predation, Profiteering and State Weakness'. London: Royal Institute of International Affairs, 2018.

El-Katiri, M. *State-building Challenges in a Post-revolution Libya*. Carlisle PA: US Army War College Strategic Studies Institute, 2012.

Gärtner, H. 'The Responsibility to Protect (R2P) and Libya', *Australian Institute for International Affairs* 1090:7, 2011.

Gaub, F. *Libya: The Struggle for Security*. Paris: European Union Institute for Security Studies, 2013.

Goulter, C. 'The UK Political Rationale for Intervention and its Consequences. The War in Libya. The Political Rationale for France', in D. Henriksen and A.K. Larssen (eds), *Political rationale and international consequences of the war in Libya*. Oxford: Oxford University Press, 2016.

Grimm, S. et al. (eds). *The Political Invention of Fragile States: The Power of Ideas*. Abingdon: Routledge, 2014.

Hameed, M. 'Responsibility to Protect: The Use and The Abuse'. New York: CUNY Academic Works, 2014.

Hobson, C. 'Responding to Failure: The Responsibility to Protect after Libya', *Millennium: Journal of International Studies* 44:3, 2016.

House of Commons Foreign Affairs Committee. *Libya: Examination of Intervention and collapse and the UK's future policy options*. London: The Stationery Office, 2016.

International Crisis Group. 'Popular Protest in North Africa and the Middle East (V): Making Sense of Libya', *Middle East/North Africa Report* 107, 2011.

International Crisis Group. 'The Prize: Fighting for Libya's Energy Wealth', *Middle East and North Africa Report* 165, 2015.

International Crisis Group 'The Libyan Political Agreement: Time for a Reset', *Middle East and North Africa Report* 170, 2016.

International Crisis Group. 'After the Showdown in Libya's Oil Crescent', *Middle East and North Africa Report* 189, 2018.

International Crisis Group. 'Making the Best of France's Libya Summit', *Middle East and North Africa Briefing* 58, 2018.

Iqbal, Z. and H. Starr, H. *State failure in the modern world*. Stanford CA: Stanford University Press, 2015.

Kim, S.K.,'Third-party intervention in civil wars: motivation, war outcomes, and post-war development'. Iowa City IA: The Graduate College of The University of Iowa, 2012.

Kometer, M.W. and S.E. Wright. *Winning in Libya: By Design Or Default?* Paris: Institut Français des Relations Internationales, 2013.

Kuperman, A.J. 'NATO's Intervention in Libya: A Humanitarian Success?', in A. Hehir and R. Murray (eds). *Libya, the Responsibility to Protect and the Future of Humanitarian Intervention.* Basingstoke: Palgrave Macmillan, 2013.

Lacher, W. and A. Al-Idrissi. 'Capital of Militias', Small Arms Survey *Briefing Paper* 2018.

Lacher, W. and P. Cole. 'Politics by other means: Conflicting interests in Libya's security sector', Small Arms Survey *Briefing Paper* 2014.

Lindström, M. and K. Zetterlund. *'Setting the stage for the military intervention in Libya': Decisions made and their implications for the EU and NATO.* Stockholm: Swedish Ministry of Defence, 2012.

Menon, R. *The conceit of humanitarian intervention.* Oxford: Oxford University Press, 2016.

Miller, P.D. 'Strategies of Statebuilding: Causes of Success and Failure in Armed International Statebuilding Campaigns by Liberal Powers', in *APSA 2010 Annual Meeting Paper.*

Mohlin, M. 'Cloak and Dagger in Libya: The Libyan Thuwar and the role of allied special forces', in K. Engelbrekt, M. Mohlin and C. Wagnsson (eds), *The NATO intervention in Libya: lessons learned from the campaign.* Abingdon: Routledge, 2013.

O'Shea, E. 'Responsibility to Protect (R2P) in Libya: ghosts of the past haunting the future', *International Human Rights Law Review* 1:1, 2012.

Odeyemi, C. 'R2P intervention, BRICS countries, and the no-fly zone measure in Libya', *Cogent Social Sciences* 2:1, 2016.

Pargeter, A. *Libya: The Rise and Fall of Qaddafi.* New Haven CT: Yale University Press, 2012.

Randall, E. 'After Qaddafi: Development and democratization in Libya', *Middle East Journal* 69:2, 2015.

Rotberg, R.I. (ed.). *When States Fail: Causes and Consequences.* Princeton NJ: Princeton University Press, 2010.

St John, R.B. *Libya: From Colony to Revolution.* London: Oneworld Publications, 2017.

Watanabe, L. 'Libya: In the Eye of the Storm', *CSS Analysis in Security Policy* 193, 2016.

Watanabe, L. *Islamist Actors: Libya and Tunisia* . Zurich: ETH, 2018.

Winer, J.M. 'Origins of the Libyan Conflict and Options for Its Resolution', *The Middle East Institute: Policy Paper 2019-4*, 2019.

Zambakari, C. 'The misguided and mismanaged intervention in Libya: Consequences for peace', *African Security Review* 25:1, 2016.

Zenko, M. 'The Big Lie About the Libyan War', *Foreign Policy*, 22 March 2016.

8

SAFE HAVENS

The Failed Strategy
of the Failed State?

Emily Knowles and Abigail Watson

Introduction

In the immediate aftermath of the 9/11 terrorist attacks on the US, then-British Foreign Secretary Jack Straw predicted the emergence of '[a] future in which unspeakable acts of evil are committed against us, coordinated from failed states in distant parts of the world'.[1] Subsequently, the strategic imperative of denying terrorist groups safe haven in fragile or failed states has been a pivotal part of the military and political rationale linking US and allied military action against groups like al-Qaeda in the Arabian Peninsula (AQAP), the Taliban, the so-called Islamic State (IS), Boko Haram and al-Shabaab to core national security concerns of preventing further attacks on their soil.[2] As General John Nicholson, the then-commander of the NATO Resolute Support mission in Afghanistan said in February 2017 evidence to the US Senate: 'Our mission was to ensure that Afghanistan

---

1   Quoted in Harry Verhoeven, 'The Self-Fulfilling Prophecy of Failed States: Somalia, State Collapse and the Global War on Terror', *Journal of Eastern African Studies* 3:3 (2009), 405-25.

2   Anthony Dworkin, *Europe's New Counter-Terror Wars* (Brussels: European Council on Foreign Relations, 2016), available at http://www.ecfr.eu/publications/summary/europes_new_counter_terror_wars7155; Brian Michael Jenkins, *Disrupting Terrorist Safe Havens*, RAND (2014), available at https://www.rand.org/blog/2014/08/disrupting-terrorist-safe-havens.html.

would never again be a safe haven for al-Qaeda or other terrorist groups to attack America or our allies and partners. That mission has been successful for 15 years, but it is not over'.[3]

Even now, following a dramatic drawdown of troops in Iraq and Afghanistan, budget cuts and a greater sense of war weariness within the British parliament and among the wider public, the UK and its partners are continuing attempts to operate in failed or failing states on a lighter footprint. In an approach referred to here as 'remote warfare', the UK engages in potential safe havens around the world, playing a largely supporting role alongside local and regional forces that do the bulk of the frontline fighting.[4] In this sense, remote warfare has allowed the British government to operate without having to reassess the utility of its safe haven-focused approach or the implications of doing it with a lighter footprint.

This chapter will argue that, while this approach has delivered some military successes, such as pushing back the territorial gains of IS, its overall impact on the health of British foreign policy has been detrimental. As examples of UK operations in Afghanistan, the anti-IS coalition and Somalia show, Western attempts to address safe havens have been reactive, short-term and have led to either further involvement or further instability. Nor has this been improved by a light footprint, which has exacerbated many of these initial problems and delayed frank discussion regarding the utility of Western strategy.[5]

Safe Havens: The Concept

There does not appear to be an exact definition of a 'safe haven' in UK doctrine and policy. However, the US State Department does define a

---

3   Gen. John Nicholson, 'Testimony on the Situation in Afghanistan', Senate Committee on Armed Services (2017), available at https://www.armed-services.senate.gov/imo/media/doc/17-08_02-09-17.pdf.
4   Emily Knowles and Abigail Watson, *Remote Warfare: Lessons Learned from Contemporary Theatres* (Oxford: Oxford Research Group, 2018), available at https://www.oxfordresearchgroup.org.uk/remote-warfare-lessons-learned-from-contemporary-theatres; Emily Knowles and Abigail Watson, *No Such Thing as a Quick Fix: The Aspiration-Capabilities Gap in British Remote Warfare* (Oxford: Oxford Research Group, 2018), available at https://www.oxfordresearchgroup.org.uk/no-such-thing-as-a-quick-fix-the-aspiration-capabilities-gap-in-british-remote-warfare.
5   Our analysis in this chapter draws on a series of expert roundtables held in London from March 2017–June 2018, our commissioned research on drone operations in the border region between Afghanistan and Pakistan, primary interviews with British military personnel in Somalia and Nigeria, and field research in Afghanistan, Iraq, Mali and Kenya.

terrorist safe haven as: 'an area of relative security exploited by terrorists to indoctrinate, recruit, coalesce, train, and regroup, as well as prepare and support their operations'.[6] For both states, denying terrorists safe havens has been presented by consecutive leaders as a way to undermine these groups' capacity to operate effectively. This is a logic that gained particular currency after 9/11.

Then-President George W. Bush said on the five-year anniversary of the attacks that a key lesson of 9/11 was that al-Qaeda needed 'a safe haven to plan and launch attacks on America and other civilised nations'.[7] Afghanistan was seen as the model for this assertion. The Soviet-Afghan war had lasted for nearly nine years and killed an estimated one million civilians, as well as tens of thousands of Mujahedeen fighters and Afghan and Soviet soldiers.[8] The 'free-for-all' created after the Soviet withdrawal in 1989 led the once-centralised Afghan state to wither away, as warlords carved up territory and dismantled formal political institutions. Eventually, this vacuum of authority was filled by a militant group – the Taliban, who provided fertile ground for Osama bin Laden and al-Qaeda to build their strength.[9] In 1996, five years before the 9/11 attacks, bin Laden made a formal declaration of 'war' against the US.[10] Significantly, UN Security Council (UNSC) Resolution 1373, a counterterrorism measure adopted unanimously following 9/11, specifically referenced terrorists' ability to move across international borders and find safe haven to solicit and move funds and to acquire weapons.[11] Similar trends can be seen in national counterterrorist strategies. In 2010, the US National Security Strategy stated: 'We will disrupt, dismantle, and defeat al-Qa'ida and its affiliates through a comprehensive strategy that denies them safe haven'.[12] This was the premise for the war in Afghanistan and for the expansion of drone

---

6   See Office of the Coordinator for Counterterrorism, 'Chapter 3 – Terrorist Safe Havens', U.S. Department of State (2006), available at http://www.state.gov/j/ct/rls/crt/2005/64333. htm.

7   George W. Bush, 'Address to the Nation on the War on Terror', 11 September 2006, *The American Presidency Project*, available at http://www.presidency.ucsb.edu/ws/?pid=73962.

8   Alan Taylor, 'The Soviet War in Afghanistan, 1979-1989', *The Atlantic* (2014), available at https://www.theatlantic.com/photo/2014/08/the-soviet-war-in-afghanistan-1979-1989/ 100786/.

9   See Verhoeven, 'The Self-Fulfilling Prophecy of Failed States'.

10 Peter Bergen, 'Al Qaeda, the Organization: A Five-Year Forecast', *The Annals of the American Academy of Political and Social Science* 618 (2008), 14-30.

11 US State Department, 'Chapter 3 – Terrorist Safe Havens'.

12 Micah Zenko and Amelia Mae Wolf, 'The Myth of the Terrorist Safe Haven', *Foreign Policy* (blog), 16 January 2015, available at https://foreignpolicy.com/2015/01/26/ al-qaeda-islamic-state-myth-of-the-terrorist-safe-haven/.

operations into Pakistan, Yemen and Somalia.[13] The Trump administration has continued to explain the benefits of overseas military engagement in these terms. In a frequently-cited conversation between then-newly elected President Donald Trump and his then-Secretary of Defense James Mattis, the latter explained that Trump needed to send more soldiers to Afghanistan 'to prevent a bomb going off in Times Square'.[14] More recently, when Trump took to Twitter to announce the withdrawal of US troops from Syria, he said: 'We have defeated ISIS in Syria, my only reason for being there'.[15]

The same beliefs have also been evident on the British side of the Atlantic. Soon after 9/11, Jack Straw stressed the need to eradicate safe havens in places like 'Somalia, Liberia and Congo [which] invoke the Hobbesian image of a "state of nature" without order, where continual fear and danger of violent death render life nasty, brutish and short'.[16] In 2013, then-Foreign Secretary William Hague noted that dealing with terrorist threats to the UK meant working 'to eliminate safe havens for it further afield'.[17] In her response to the London Bridge terrorist attacks in 2017, then-Prime Minister Theresa May stated that there needed to be a greater effort to disrupt safe havens, calling on her government to 'not forget about the safe spaces that continue to exist in the real world. Yes, that means taking military action to destroy ISIS in Iraq and Syria'.[18]

Remote Warfare: The Contemporary British Approach

The failure of two costly military interventions in Iraq and Afghanistan to establish expected levels of stability has led some commentators to

---

13 Ibid.

14 Quoted in Trevor Thrall and Erik Goepner, 'Policy Roundtable: 17 Years After September 11', *Texas National Security Review* (2018), available at https://tnsr.org/roundtable/policy-roundtable-17-years-after-september-11/.

15 Amanda Sakuma, 'Trump's Syria Withdrawal Now Comes with Conditions, Bolton Says', *Vox*, 6 January 2019, available at https://www.vox.com/2019/1/6/18170788/john-bolton-syria-withdrawal-conditions.

16 Verhoeven, 'The Self-Fulfilling Prophecy of Failed States'.

17 William Hague, 'Countering Terrorism Overseas', *GOV.UK*, 14 February 2013, available at https://www.gov.uk/government/speeches/countering-terrorism-overseas.

18 See 'Britain's Prime Minister Theresa May's Full Speech on the London Attacks', *Time*, 4 June 2017, available at http://time.com/4804640/london-attack-theresa-may-speech-transcript-full/.

Unilateral operations (us)	Partner operations (us + them)	Train, advise, assist (them + us)	Security assistance (them)
Targeted killing	'Mentor' roles	Training local troops/armed groups	Arms sales/transfers
Limited strikes	Joint (blue/green) operations	Advising (ministerial or planning)	Political support
	'Accompany' roles		Financial support
	Targeting support (JTAC/ISR)		
Local forces 'shadowing' UK operations			
	Air support		

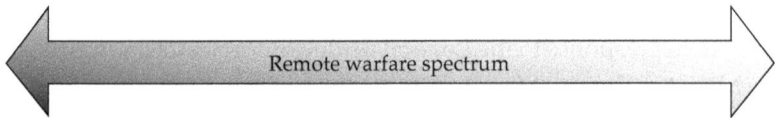

Remote warfare spectrum

Figure 8.1: Rethinking Military Intervention Abroad

announce the 'death of the nation-building project'.[19] Placing comparable numbers of Western boots on the ground again, except in the case of a direct threat to state survival, does not seem likely – at least for another generation, when memories and national budgets may have recovered.[20]

For the UK, among others, this has meant rethinking approaches to military intervention abroad (the logic behind such an approach is summarised in Figure 8.1 above). NATO commitments in places like Afghanistan have been reduced to 10,000 troops from a high of 100,000, and priorities in many European states have been refocused.

A resurgent Russia has seen a number of Western militaries, including the UK and the US, reposition themselves for state-on-state conflicts

---

19 See, for example, Doug Bandow, 'The Nation-Building Experiment That Failed: Time For U.S. To Leave Afghanistan', *Forbes*, 1 March 2017, available at https://www.forbes.com/sites/dougbandow/2017/03/01/the-nation-building-experiment-that-failed-time-for-u-s-to-leave-afghanistan/#35f9a24c65b2.

20 Rachel Gribble et al, 'British Public Opinion after a Decade of War: Attitudes to Iraq and Afghanistan', *Political Studies Association* (2014), available at https://www.kcl.ac.uk/kcmhr/publications/assetfiles/2014/Gribble2014b.pdf; 'David Cameron: 'Syria Is Not like Iraq'', *BBC News*, available at http://www.bbc.co.uk/news/uk-politics-23883970.

– to the detriment of the debate on current operations. For instance, one soldier deployed to Afghanistan in March 2017 recounted how people at home almost did not believe it when told that they had been sent out to Afghanistan – believing that NATO's operations in the country 'were done'.[21]

Nevertheless, governments continue to worry about the impact of terrorist activity, thriving in the world's ungoverned or weakly-governed spaces, on their own national security. In order to deny terrorist groups safe havens, some unilateral counterterrorism strikes and raids continue – like the strike against IS propagandist and British citizen Reyaad Khan, who was killed in Syria in August 2015,[22] or the dropping of the 'mother of all bombs' (MOAB) on IS positions in Afghanistan in April 2017.[23] The exploitation of Western technological superiority – particularly from the air – has allowed states like the UK to engage in the fight against groups like IS without putting large numbers of their own boots on the ground.

This is perhaps the most visible aspect of what is termed here as 'remote warfare'. However, Western troops are also increasingly working with and through local and regional allies in important areas for global security. Through this approach, local troops are expected to do the bulk of the frontline fighting against groups like Boko Haram, al-Qaeda, IS and al-Shabaab, while small teams of Special Forces (SF) and military advisers, as well as security assistance and intelligence support, are often provided by Western partners, as noted in Figure 8.2 below.[24]

By maintaining a light footprint, some of the political and military risks of exposing British troops to another series of gruelling wars appear to have been minimised. From a domestic perspective, there have been no high-profile anti-war protests on the streets of London similar to the million people claimed to have marched in the lead up to the Iraq War and – bar the embarrassing defeat in parliament of a government motion on the principle

---

21 Author interviews in Afghanistan, March 2017.
22 Shiv Malik et al, 'Ruhul Amin and Reyaad Khan: The Footballer and the Boy Who Wanted to Be First Asian PM', *The Guardian*, 7 September 2015, available at https://www.theguardian.com/world/2015/sep/07/british-isis-militants-killed-raf-drone-strike-syria-reyaad-khan-ruhul-amin.
23 Raf Sanchez and Charlotte Krol, 'Afghanistan Blast: BBC Driver Killed and Four Journalists Injured after Huge Car Bomb Kills at Least 80 in Kabul's Diplomatic Quarter', *The Daily Telegraph*, 31 May 2017, available at http://www.telegraph.co.uk/news/2017/05/31/afghanistan-huge-explosion-near-presidential-palace-kabul/.
24 See Emily Knowles and Abigail Watson, 'How Can We Win? Lessons Learned from Contemporary Theatres', *Agile Warrior Quarterly* (2018), available at https://www.oxfordresearchgroup.org.uk/agile-warrior-quarterly.

The ideological and budgetary death of the nation-building project
+
Technological innovation
+
Low public and parliamentary trust
+
High political risk aversion
+
The perceived security threat of safe havens
+
The weakness of local partners
=
A strong incentive for the UK to engage discretely and without putting large numbers of their *own* boots on the ground.

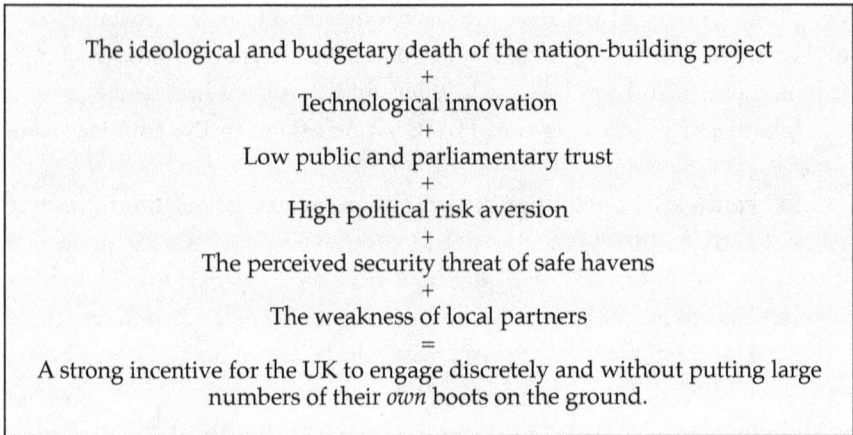

Figure 8.2: The Spectrum of Remote Warfare

of military action in Syria in 2013[25] – the UK has been able to lend support to its allies relatively unhindered, despite a seemingly war-weary public and parliament. The high-profile liberations of Mosul and Raqqa from IS control have done much to reassure critics that this model of engagement can enable states like the UK to push back terrorist groups with little risk to their own forces and that, with the right support, local fighters can prevail.

As such, key players within the wider international community are using remote warfare to remain engaged against terrorist groups in places like Afghanistan, Iraq, Somalia, Libya, Yemen, Nigeria and Mali – even as large-scale Western boots on the ground-type operations remain unpalatable to the politicians and publics back at home.[26] However, as this chapter will explore, the limited success of these operations raises vital questions about how useful a safe-haven driven approach to countering terrorist networks is proving to be.

Three cases will be examined in more depth here, namely Afghanistan, the anti-IS coalition in the Middle East and Somalia. In each

25 Abigail Watson, 'Pacifism or Pragmatism? The 2013 Parliamentary Vote on Military Action in Syria', *Remote Warfare Programme* (blog), 29 May 2018, available at https://www.oxfordresearchgroup.org.uk/pacifism-or-pragmatism-the-2013-parliamentary-vote-on-military-action-in-syria.
26 Emily Knowles and Abigail Watson, 'All Quiet On The ISIS Front: British Secret Warfare In The Information Age', *Remote Warfare Programme* (blog), 31 March 2017, available at https://www.oxfordresearchgroup.org.uk/all-quiet-on-the-isis-front-british-secret-warfare-in-an-information-age.

case, remote warfare has allowed the UK to respond to potential terrorist threats emanating from safe havens with a light military footprint. This has minimised pressure on defence budgets, domestic public opinion and from parliamentary inquiries. However, in doing so it has focused its activities on short-term military objectives in regions where there are widespread and complex political problems. Even when this approach has delivered some military success – such as pushing IS back from the territory it controlled in Iraq – it has also shown signs of undermining long-term prospects of peace and stability.

Afghanistan

Over 17 years after the US entered Afghanistan, the war-weary state is showing signs of returning to its pre-September 11 status as a safe haven for terrorist groups. One Afghan scholar noted in 2013 that:

> In spite of some 150,000 well-equipped foreign troops and over 300,000 Afghan military and police, security is deteriorating in Afghanistan. People are not safe in their cities and villages, the government is ineffective, judges, police and government bureaucrats are corrupt and above all Pakistan and Iran are sabotaging whatever the coalition forces are building.[27]

Since 2014, the ongoing NATO mission has been reduced to a notably lighter footprint of 15,000 personnel under a largely Train, Advise and Assist (TAA) mandate to consolidate the skills of the Afghan National Defence and Security Forces (ANDSF), with an eye to gradually handing over tasks and capabilities to local troops. This is a remote form of engagement, with frontline fighting predominantly engaged in by the local troops in the ANDSF, with Western training, advice, support and a small number of SF on the ground to accompany troops. While there may be a certain *national* logic to such an approach, the problem is that, even after all this time, Afghanistan does not seem ready for such a notable shift in responsibilities.

Western counterterrorism professionals in South Asia have noted that al-Qaeda is rebuilding itself because it has been given the space to do so by a weak domestic security sector. Afghanistan's security forces lack resources

---

27 Alam Payind, 'Inside Afghanistan 23 Years after the Soviet Withdrawal', *Journal of Asian and African Studies* 48:2 (2013), 258–64.

and training and face growing attrition, due to the deteriorating national environment, as depicted by a fragile economy and weak political system, and a failure on the part of external actors to continue to prioritise it within their wider foreign policies.[28] Interviews undertaken in March 2017 found many ordinary soldiers that agreed with this diagnosis, with one declaring that they had been deployed on a 'forgotten mission'.[29]

Signs of a terrorist resurgence came as early as October 2015, when a joint US-Afghan SF operation uncovered the largest al-Qaeda camp ever found in the region. The ensuing multi-day battle killed more than 160 jihadist fighters in a training camp facility that spanned 30 square miles.[30] More recently, the Special Inspector General for Afghanistan Reconstruction (SIGAR) was asked not to release the usual figures on Taliban and IS control of Afghan territory in their January 2018 report, assumedly because they did not make for positive reading.[31] However, a BBC study during the same month estimated that Taliban fighters were openly active in 70 per cent of the country. It found that the Taliban were in full control of 14 districts and 'have an active and open physical presence in a further 263'.[32]

There were signs of movement during 2018, when President Trump increased the number of US troops in theatre and asked other states to do the same. The UK government stated that Afghanistan was an 'enduring' commitment and sent an 'additional 440 personnel in non-combat roles to take the total UK contribution to around 1,100 personnel'.[33] However, during field research carried out by the Oxford Research Group's Remote Warfare Programme team in March 2017, there was a clear sense that additional troops would do little to improve a poor strategy. When one soldier was asked what they thought to having more NATO troops on the ground in Afghanistan, they paused and then shrugged: 'I'm not sure that wouldn't just make us a bigger target'. Another reflected that military personnel were currently just 'filling the vacuum left by other actors and agencies' – there

---

28 See Sara A. Carter, 'Afghanistan Quickly Becoming Safe-Haven for Terror Groups, Again', *Circa*, 15 September 2016, available at http://circa.com/politics/afghanistan-quickly-becoming-safe-haven-for-terror-groups-again.
29 Author interviews in Afghanistan, March 2017.
30 Carter, 'Afghanistan Quickly Becoming Safe-Haven for Terror Groups, Again'.
31 SIGAR, 'Quarterly Report to Congress: Section 3, Security', 30 January 2018, available at https://www.sigar.mil/pdf/quarterlyreports/2018-01-30qr-section3-security.pdf.
32 Shoaib Sharifi and Louise Adamou, 'Taliban 'Threaten 70% of Afghanistan'', *BBC News*, 31 January 2018, available at http://www.bbc.co.uk/news/world-asia-42863116.
33 Gavin Williamson, 'Afghanistan', *Hansard Debate*, 11 July 2018, available at https://hansard.parliament.uk/Commons/2018-07-11/debates/9AA19D37-6365-43BC-910A-A78461549C8D/Afghanistan.

is a danger, then, of more troops doing the same. The political imperative to 'end' NATO operations in the country compounded problems that can be seen vividly on the ground today. As another soldier put it, 'we went too far and shed capacities that we needed – things like counter narcotics, countercorruption, the ability to trace money through the Afghan system'.[34]

The pressures of delivering on security and breaking the 'stalemate'[35] in Afghanistan were keenly felt by the soldiers on the ground.[36] As one remarked 'we face a stalemate today, but we also faced one 5, 8, 10, 15 years ago, we just didn't know it'. For many, it was clear that, as there were problems maintaining an 'acceptable' level of security in Afghanistan even with tens of thousands of troops on the ground, rectifying them on the basis of a significantly lighter footprint was an impossible task. Soldiers repeatedly remarked that 'ideally any force should have its size based on the conditions on the ground, and the end you are trying to achieve', not on domestic political constraints and shifts in international attention.[37]

Added to this, high risk-aversion has meant that states are adopting restrictive Rules of Engagement (RoE), with troops being asked to conduct effective training of local security forces while being essentially constrained by their HQs. For many soldiers, this further exacerbated the problem, with one interviewee lamenting that 'this is making relationship-building really hard. We can't go out and interact like we used to', while another noted that 'we know that if anyone gets killed, we could lose our strategic freedom of action'.[38]

There is a general appreciation that the problems in Afghanistan are long-term, with some personnel suggesting that there is likely to be some sort of international TAA mission there indefinitely. As one international soldier put it, '[Afghanistan] is the US's longest war. It is a war. This puts

---

34 Author interviews in Afghanistan, March 2017.

35 'Top US General: The War in Afghanistan Is Now a 'Stalemate'', *Business Insider*, 9 March 2017, available at http://www.businessinsider.com/top-us-general-the-war-in-afghanistan-stalemate-2017-3; 'U.S. Marines Back in Helmand as Afghanistan 'stalemate' Continues', *Reuters*, 1 May 2017, available at http://www.reuters.com/article/us-afghanistan-marines-idUSKBN17W07L; Michael R. Gordon, 'Trump Advisers Call for More Troops to Break Afghan Deadlock', *The New York Times*, 8 May 2017, available at https://www.nytimes.com/2017/05/08/us/politics/donald-trump-afghanistan-troops-taliban-stalemate.html.

36 Andrew Shaver and Joshua Madrigal, 'Losing in Afghanistan', *Foreign Affairs* 22 September 2016, available at https://www.foreignaffairs.com/articles/afghanistan/2016-09-22/losing-afghanistan; Sabera Azizi, 'America Can't Afford to Keep Losing the War in Afghanistan', *The National Interest*, 11 April 2017, available at http://nationalinterest.org/blog/the-skeptics/america-cant-afford-keep-losing-the-war-afghanistan-20127.

37 Author interviews in Afghanistan, March 2017.

38 Ibid.

us into the mental construct of it having a beginning and an end ... But look at Korea, look at our commitments to West Germany, we were there for 70 years. But politicians don't want to hear that'.[39] While there may not prove to be temporal limits to some level of commitment, operations in Afghanistan have shown that there are territorial limits to a safe-havens approach to warfare, with even powerful actors, such as the US, unable to pursue hostile groups wherever they may be found. Many commentators agree that shifting the focus to continued safe havens in Pakistan has undermined military progress in Afghanistan.[40] As Bruce Hoffman notes: 'Every day that the US allows the unsatisfactory situation along both borders to continue is another day that al-Qaeda and its allies have to regroup, reorganize, and marshal their strength'.[41] Even where the US has extended its strike operations into Pakistani territory, there is little evidence that this has been a successful deterrent to terrorist groups. An Oxford Research Group report commissioned in 2014 into drone strikes in the Federally Administered Tribal Areas (FATA) of Pakistan found that, as well as disrupting terrorist activity and killing some high-value targets, drone strikes have also exacerbated the spread of militants across the country. Large numbers of terrorists have fled from the heavily targeted FATA region 'into the settled and less restive parts of the country, creating problems for their new host communities'. This has had a number of negative implications for wider Pakistani society, including increased radicalisation, sectarian and gang violence and drug and weapon smuggling.[42]

In order to really begin to make progress on a peaceful settlement for Afghanistan, interviewees spoke of the need to bring pressure to bear on states such as Pakistan, Russia and the United Arab Emirates (UAE) to restrict assistance to the Taliban, to start supporting the delivery of a functioning economy alongside the provision of security and to build the trust necessary between NATO troops and their Afghan counterparts so

---

39 Ibid.

40 Payind, 'Inside Afghanistan 23 Years after the Soviet Withdrawal'; Peter Bergen and Laurence Footer, 'Defeating the Attempted Global Jihadist Insurgency: Forty Steps for the Next President to Pursue against Al Qaeda, Like-Minded Groups, Unhelpful State Actors, and Radicalized Sympathizers', *The ANNALS of the American Academy of Political and Social Science* 618:1 (2008), 232-47.

41 Bruce Hoffman, 'A Counterterrorism Strategy for the Obama Administration', *Terrorism and Political Violence* 21:3 (2009), 359-77.

42 See Wali Aslam, *Terrorist Relocation and the Societal Consequences of US Drone Strikes in Pakistan* (Oxford: Oxford Research Group 2014), available at https://www.oxfordresearchgroup. uk/drones-in-pakistan-relocating-terrorists-not-eliminating-them.

that more roles and responsibilities can be handed over.[43] Similarly, while recent peace talks are a positive step forward, they have also indicated some of the shortcomings of Western efforts; especially given that, even after 18 years of building up the Afghan national government, the Taliban still insist on bypassing it in negotiations, preferring to deal directly with the US and 'prominent Afghan figures'.[44] None of these problems will be rectified by remote warfare alone, especially if the focus remains primarily on countering safe havens militarily – to the detriment of economic, developmental and political levers of power.

The Anti-IS Coalition

David Ignatius, a journalist at the *Washington Post*, noted that the rise of IS 'teaches the same basic lesson that emerged from America's other failures in the Middle East over the last decade', specifically '[a]ttempts … to topple authoritarian regimes … create power vacuums. This empty political space will be filled by extremists unless the United States and its allies build strong local forces that can suppress terrorist groups and warlords'.[45] In January 2014, three years after the formal withdrawal of international military forces from Iraq, a hitherto little-known group calling itself Islamic State in Iraq and Syria sprang onto the international stage when it seized the Iraqi city of Fallujah, which sits just 43 miles west of Baghdad. A few weeks later, IS stormed into the Syrian city of Raqqa and announced it as the group's 'capital'. By the end of June 2014, Mosul and Tikrit had also fallen and IS had declared the establishment of a caliphate, naming its leader Abu Bakr al-Baghdadi as Caliph and successor to the Prophet Mohammed.

In response, a US-led anti-IS coalition conducted airstrikes and built the capacity of some of the local forces already fighting IS. This included the Kurdish-dominated People's Protection Units (Yekîneyên Parastina Gel or YPG), which later expanded into the Syrian Democratic Forces (SDF), and the Peshmerga, the military forces of the federal region of Iraqi Kurdistan. These operations had some success in dramatically cutting the

---

43 Author interviews in Afghanistan, March 2017.
44 Kathy Gannon and Rahim Faiez, 'Taliban to Take Part in 'Intra-Afghan' Talks in Moscow', *The Diplomat*, 5 February 2019, available at https://thediplomat.com/2019/02/taliban-to-take-part-in-intra-afghan-talks-in-moscow/.
45 David Ignatius, 'How ISIS Spread in the Middle East', *The Atlantic*, 29 October 2015, available at https://www.theatlantic.com/international/archive/2015/10/how-isis-started-syria-iraq/412042/.

territorial gains of IS.[46] The Iraqi government announced the liberation of Mosul in July 2017 and the SDF declared Raqqa liberated in October 2017.[47] Unfortunately, in focusing on removing a terrorist safe haven from Iraq and Syria, the US, UK and other allies have attempted to engage in broader conflicts within a narrow counterterrorism mandate. This has empowered groups that may impede the broader stability needed to provide a long-term solution to the emergence of terrorist groups in the region.[48]

In Syria, the SDF has proved a crucial partner in pushing back IS. The group began life as the YPG, which started fighting IS when its militants moved to take the Syrian Kurdish town of Kobani in September 2014. The international anti-IS coalition provided air support to YPG forces, defeating IS in the area. As one commentator notes, 'an operational relationship was born. The midwife was tactical necessity. Larger issues of national security objectives, overall strategy for Syria, and an important bilateral relationship with a NATO partner were made subordinate to the singular focus on attacking ISIS'.[49] While this relationship was successful in pushing back IS and re-establishing peace and stability in majority-Kurdish areas, once the SDF moved into other parts of the country it served to increase tensions between Kurds and Arabs. While the SDF made efforts to diversify its ranks and include a notable number of Arab recruits, 'in practice it remains squarely under YPG command'.[50] As Haid, of Chatham House and the

---

46 By mid-2014 the group controlled a taxable population of some seven or eight million, oilfields and refineries, vast grain stores, lucrative smuggling routes and vast stockpiles of arms and ammunition, as well as entire parks of powerful modern military hardware. This has now been almost completely retaken.

47 Jack Detsch, 'Raqqa Win Pulls US Deeper into Syria Conflict', *Al-Monitor*, 17 October 2017, available at http://www.al-monitor.com/pulse/originals/2017/10/raqqa-win-syria-us-deeper-into-conflict-ypg.html; Barack Barfi, 'The Battle for Raqqa – It's Complicated', *The Cipher Brief*, 31 March 2017, available at https://www.thecipherbrief.com/article/middle-east/battle-raqqa-its-complicated-1091; Bethan McKernan, 'Isis Is Regrouping for Battle after Losing Mosul and Raqqa, Warn Libyan Forces', *The Independent*, 27 July 2017, available at http://www.independent.co.uk/news/world/middle-east/isis-regrouping-libya-forces-mosul-raqqa-islamic-state-syria-iraq-islamic-state-a7862606.html; Renad Mansour and Erwin Van Veen, 'Iraq's Competing Security Forces After the Battle for Mosul', *War on the Rocks*, 25 August 2017, available at https://warontherocks.com/2017/08/iraqs-competing-security-forces-after-the-battle-for-mosul/.

48 Robert Malley, 'What Comes After IS?', *Foreign Policy* (blog), 10 July 2017, available at https://foreignpolicy.com/2017/07/10/what-comes-after-IS-islamic-state-mosul-iraq-syria/.

49 Aaron Stein, 'Partner Operations in Syria: Lessons Learned and the Way Forward', Atlantic Council of the US, July 2017, available at http://www.atlanticcouncil.org/publications/reports/partner-operations-in-syria.

50 'Fighting ISIS: The Road to and beyond Raqqa', *Crisis Group*, 28 April 2017, available at https://www.crisisgroup.org/middle-east-north-africa/eastern-mediterranean/syria/b053-fighting-isis-road-and-beyond-raqqa.

International Centre for the Study of Radicalisation and Political Violence, notes, 'many Syrian Arabs saw the SDF's attack on rebel-held areas ... as a Kurdish pretext to take advantage of US support and expand their territories in areas where Arabs are a majority'.[51] The SDF and its political affiliate, the Syrian Democratic Council, (SDC) set up bureaucratic structures in liberated territory at record speed and now controls large swathes of territory in north-eastern Syria. However, Crisis Group argues that '[o]utside majority-Kurdish areas ... [the YPG] governance model appears fragile'.[52] They describe efforts 'to achieve Arab buy-in to its project' as 'partial and haphazard' rather than 'meaningful'.[53] Haid goes further, stating that 'reported violations committed by some Kurdish groups against Arab communities have led to ethnic tensions between local communities'.[54]

In Iraq, the UK and its allies have given substantial support to the Peshmerga to help them push back IS.[55] As described by a representative from the Foreign and Commonwealth Office (FCO):

> As part of the package of assistance provided by the Global Coalition to counter Daesh, we have provided the Peshmerga with military support ... UK training teams have trained over 57,000 members of the Iraqi Security Forces, including 9,000 Peshmerga fighters; since September 2014 we have gifted £3 million of arms and ammunition to the Peshmerga; and the UK has given air support to the Peshmerga as part of the Coalition.[56]

Many high-level officials – engaged with during research by the Remote Warfare Programme – complained there was insufficient focus on the post-IS phase throughout the campaign, including what the long-term implications of supporting the Peshmerga might be.[57] In particular, one consequence has been to increase Peshmerga capabilities vis-à-vis the Iraqi state, prompting the Kurdish regional government to hold an independence referendum in

51 Haid Haid, 'The Deal between Syrian Arabs and Kurds That Could Be the Way Forward in Northern Syria', *Middle East Eye*, 15 May 2017, at http://www.middleeasteye.net/columns/deal-between-syrian-arabs-and-kurds-could-be-way-forward-northern-syria-274041731.

52 Crisis Group, 'Fighting ISIS'.

53 Ibid.

54 Haid Haid, 'The West's Obsession with IS Has Not Helped Syria', *CNN*, 7 April 2017, available at https://www.cnn.com/2017/03/15/opinions/strategy-syria-mess-haid-haid-opinion/index.html.

55 'Iraqi Kurds 'Fully Control Kirkuk'', *BBC News*, 12 June 2014, available at https://www.bbc.co.uk/news/world-middle-east-27809051.

56 House of Commons Foreign Affairs Select Committee, *Kurdish Aspirations and the Interests of the UK* (London: The Stationery Office, 2018), available at https://publications.parliament.uk/pa/cm201719/cmselect/cmfaff/518/51802.htm.

57 'Iraqi Kurds 'Fully Control Kirkuk''.

Kurdistan and some disputed territories, such as Kirkuk, which they had taken back from IS in September 2014. In response, Iraqi forces retook Kirkuk, reversing any goodwill that had been forged during the fight against IS and splitting the nascent Kurdish unity.[58] This also raised doubts among the Peshmerga over whether Western support would exist after the fall of IS.[59]

Another complicating factor for the reunification of the Iraqi security sector has been the rise of the Popular Mobilisation Forces (*al-hashd al-shaabi*, or PMF) – a conglomeration of predominantly Shia Iraqi militias formed during the early stages of the anti-IS campaign. A senior British military officer claimed the UK would not conduct strikes in direct support of the PMF due to the Iranian links of some of their units; however, Major General Rupert Jones (Deputy Commander of the Combined Joint Task Force for Operation Inherent Resolve) acknowledged that the PMF 'incidentally' benefited from Western 'support provided to Iraqi security forces under the control of Baghdad'.[60] This has contributed to claims the Iraqi army 'is lucky if it can be considered the fourth-strongest army in Iraq – behind, Kurdistan's Peshmerga forces, the ... PMF, ... and Iraqi tribal fighters'.[61] This security sector fragmentation cannot help but undermine efforts to rid the region of extremist groups. As Frank Archibald, the former Director of the Central Intelligence Agency's (CIA) National Clandestine Service, states:

> Any counterterrorism success of clearing a city or a village of IS ... will be undercut if there is not a government of the day to do the hard work of governance the day after military success ... To fail at this ... creates an ungoverned space that will significantly enable terrorism, crime and human suffering.[62]

Remote warfare will not provide the solutions to these problems, nor will attempting to rid Iraq and Syria of safe havens through military means

---

58 Loveday Morris, 'How the Kurdish Independence Referendum Backfired Spectacularly', *Washington Post*, 20 October 2017, available at https://www.washingtonpost.com/world/how-the-kurdish-independence-referendum-backfired-/2017/10/20/3010c820-b371-11e7-9b93-b97043e57a22_story.html.

59 Knowles and Watson, *No Such Thing as a Quick Fix*.

60 Barbara Opall-Rome, 'Commander: US-Led Assault in Iraq 'Incidentally' Benefits Iran-Backed Militias', *Defense News*, 25 August 2017, available at https://www.defensenews.com/global/mideast-africa/2017/08/25/commander-us-led-assault-in-iraq-incidentally-benefits-iran-backed-militias/.

61 Renad Mansour and Faleh Jabar, 'The Popular Mobilization Forces and Iraq's Future', *Carnegie Middle East Center*, 28 April 2017, available at https://carnegie-mec.org/2017/04/28/popular-mobilization-forces-and-iraq-s-future-pub-68810.

62 Bennett Seftel, 'ISIS Festers and Grows in Lawless Libya', *The Cipher Brief* (blog), 26 January 2018, available at https://www.thecipherbrief.com/isis-festers-grows-lawless-libya.

alone. Instead these states require a broader strategy, which incorporates economic, diplomatic and developmental efforts in building lasting security.

Somalia

Somalia has topped the Fragile States Index (FSI) for eight of the past 10 years.[63] Over 20 years of conflict and a history of fractious relationships between the semi-autonomous federal member entities has left the Federal Government with any degree of effective control over less than half of the country. By the end of 2017, around 20 per cent of the country was estimated to be under the control of the terrorist group al-Shabaab.[64] This group are also responsible for high-profile attacks, such as the siege at the Westgate shopping mall in neighbouring Kenya in 2013 and a double bombing in Mogadishu at the end of October 2017.[65]

There has been increased international engagement in theatre since 9/11, when – worried that al-Qaeda would use Somalia as a safe haven after operations began in Afghanistan – the US sent a small team of Special Operations Forces (USSOF) to the country, liaising with local forces in a similar model to the early days of the Afghan conflict.[66] Like the US, much of the UK's support has been on a light footprint. As early as December 2001, it was reported that the UK had been asked to help its American allies to prepare counterterrorism strikes in the country.[67] This relationship appears

---

63 Mary Harper, 'How Do You Solve a Problem like Somalia?', *BBC News*, 11 May 2017, available at https://www.bbc.com/news/world-africa-39855735.

64 Shawn Snow, 'AMISOM Withdrawal Tests U.S. Mission in Somalia', *Military Times*, 9 November 2017, available at http://www.militarytimes.com/flashpoints/2017/11/09/amisom-withdrawal-tests-us-mission-in-somalia/.

65 Daniel Howden, 'Terror in Nairobi: The Full Story behind Al-Shabaab's Mall Attack', *The Guardian*, 4 October 2013, available at http://www.theguardian.com/world/2013/oct/04/westgate-mall-attacks-kenya; 'Deadly Double Bombing Strikes Mogadishu', *Al-Jazeera*, 29 October 2017, available at http://www.aljazeera.com/news/2017/10/double-car-bombing-strikes-mogadishu-171028152255706.html.

66 Michael Smith, 'US Special Units 'Are Already at Work in Somalia'', *The Daily Telegraph*, 12 December 2001, available at http://www.telegraph.co.uk/news/worldnews/northamerica/usa/1365115/US-special-units-are-already-at-work-in-Somalia.html.

67 Robert Fox and Jessica Berry, 'Britain Asked to Prepare Strikes against Terror Bases in Somalia', *BBC News*, 2 December 2001, available at http://www.telegraph.co.uk/news/worldnews/africaandindianocean/somalia/1364106/Britain-asked-to-prepare-strikes-against-terror-bases-in-Somalia.html.

to have expanded, with a number of reports of UKSF on the ground.[68] The UK has also provided bilateral support to the Somali National Army (SNA) and the African Union Mission in Somalia (AMISOM), as well as engaging through the UN Mission in Somalia (UNSOM) and the EU Training Mission in Somalia (EUTM-S). The UK also supports African troop-contributing countries deploying to Somalia through the British Peace Support Team (Africa).[69]

Despite concerted international attention, neither the SNA nor AMISOM has been able to dislodge terrorist groups with any permanent effect. US operations also appear to be stepping up rather than winding down. In 2017, the total number of approximately 45 US drone strikes exceeded the *cumulative* number of attacks over the previous 15 years.[70] However, at the same time, AMISOM has begun to withdraw its own troops from the country.[71] Budget pressures,[72] including some disquiet over the disproportionate risks borne by regional troops relative to their international backers,[73] appear to be taking their toll.

Throughout the history of the campaign, it seems that many of these problems have often been rooted in a lack of sufficient, or consistent, political will and commitment from Western states. This was certainly the opinion of a number of British soldiers rotating out of the country; many were worried about the 'limited ability to maintain budget and interest over the long-term'.[74] One said the 'biggest threat to UK operations in Africa is political vacillation'.[75] With political will derided as 'a yo-yo',[76]

---

68 Patrick Williams, 'SAS 'fighting Secret War' against Islamic Extremists in Somalia', *Dailystar. co.uk*, 24 April 2016, available at http://www.dailystar.co.uk/news/latest-news/510296/SAS-fighting-secret-war-islamic-extremists-Somalia-covert-operation-drone-strikes.

69 See 'Deployments: Africa', available at https://www.army.mod.uk/deployments/africa/.

70 Nick Turse, 'From Afghanistan to Somalia, Special Operations Achieve Less With More', *War Is Boring* (blog), 10 January 2018, available at https://warisboring.com/from-afghanistan-to-somalia-special-operations-achieve-less-with-more/; Jack Serle and Jessica Purkiss, 'Drone Wars: The Full Data', The Bureau of Investigative Journalism, 1 January 2017, available at https://www.thebureauinvestigates.com/stories/2017-01-01/drone-wars-the-full-data.

71 Snow, 'AMISOM Withdrawal Tests U.S. Mission in Somalia'; 'AMISOM Says 1,000 Troops to Leave Somalia', *VOA*, available at https://www.voanews.com/a/african-union-force-begins-withdrawal-from-somalia/4104674.html.

72 'Security Fears in EA as Amisom Starts Withdrawing from Somalia', *The East African*, 23 December 2017, available at http://www.theeastafrican.co.ke/news/Security-fears-in-EA-as-Amisom-starts-withdrawing-from-Somalia/2558-4239576-12yuv3g/index.html.

73 'Countdown to AMISOM Withdrawal: Is Somalia Ready?', *IRIN*, 28 February 2017, available at https://www.irinnews.org/analysis/2017/02/28/countdown-amisom-withdrawal-somalia-ready.

74 Author interviews in Somalia, February 2017.

75 Author interviews in Kenya, September 2018.

76 Author interviews in Somalia, February 2017.

another soldier called the operation 'a waste of time' because 'you are either all in or you are not in at all'.[77] As another put it, it cannot be 'tap on, tap off'[78] without handing space to groups like al-Shabaab to grow and exploit the chaos. Many noted the fact that appetites tended to wane 'if immediate improvements aren't seen'. One explained that, while 'everyone wants things to happen quicker than they can do ... you have to take very small steps in order to achieve something big and significant'.[79] However, the soldiers felt that this was often not appreciated by decision makers in Whitehall. This is not a problem restricted to remote warfare – changeable political will and the prioritisation of 'quick wins' are a recurring theme in many analyses of modern counterterrorism operations.[80] However, the light-footprint nature of the British presence in Somalia was cited as making the disadvantages particularly acute. The ability of decision makers to 'throw a few soldiers here and a few soldiers there' allowed them to show that the UK was present and 'doing something' to help its partners without properly assessing its overall strategy or effect. In fact, many soldiers commented that their deployment had been to 'send a political statement' to allies rather than to improve partner capacity or move towards a more peaceful end state for the country itself.[81]

This may, in part, explain why, despite sustained Western engagement, the long-term prospects for security in Somalia are not looking particularly promising. There remain real concerns about the viability – and the acceptability – of the SNA as a long-term security provider in the country. As one soldier remarked, they are 'just another militia, albeit an apparently legitimate militia'. When they run out of ammunition, there are no procedures in place to resupply them. In many cases, there are no funds to pay them either. This results in what another soldier described as 'a big recruitment tool for al-Shabab because ... they steal, rape ... Same as others, but this time in uniform, with Somali flags on it'.[82] At the same time, al-Shabaab's attack on the DusitD2 hotel in Nairobi, Kenya, in January 2019 – which left

77 Ibid., October 2016.
78 Ibid., November 2017.
79 Author interviews in Somalia, February 2017.
80 Philip Lohaus, 'A Missing Shade of Gray: Political Will and Waging Something Short of War', *War on the Rocks*, 11 January 2017, available at https://warontherocks.com/2017/01/a-missing-shade-of-gray-political-will-and-waging-something-short-of-war/;   Thomas Rid and Thomas Keaney (eds), *Understanding Counterinsurgency: Doctrine, Operations, and Challenges* (London: Routledge, 2010), 174.
81 Author interviews in Kenya, September 2018.
82 Author interviews in Somalia, October 2016.

21 people dead – was another shocking reminder of the continuing terrorist threat posed by the group to the rest of the region.[83]

As AMISOM troops begin to withdraw, the prospect of Somalia receiving the levels – and consistency – of political attention it needs are ever more remote. Once again, attempting to address a safe haven on a limited military footprint ignores the larger institutional problems, such as the legitimacy of Somalia's own security forces. In such a climate, as efforts to degrade the terrorist threat through airstrikes continue (and even increase), the chances of improving peace and stability in the region seem ever diminishing.

Conclusion

On the 17th anniversary of 9/11, Trevor Thrall of George Mason University, and Erik Goepner of the Cato Institute, argued that 'the safe haven fallacy is an argument for endless war based on unwarranted assumptions'.[84] It certainly seems that, in Afghanistan, Iraq, Syria and Somalia, a safe haven approach has delivered little more than short-term solutions – and has often even failed to do this. Similarly, the use of remote warfare by Western militaries has often exacerbated tendencies to short-term thinking and delayed frank discussions about whether operations are working.

This is not to say there have not been some successes. Certainly, al-Qaeda has not conducted another attack against the US since 9/11. In fact, papers found at bin Laden's compound in 2011 suggested that US operations against the group had crippled its ability to operate.[85] However, at the same time, al-Qaeda has also evolved from 'a few hundred members, almost all of them based in a single country' to one that 'enjoys multiple safe havens across the world'.[86] Conflict in the Middle East has created fertile ground for groups like al-Qaeda to re-establish themselves. Former Federal Bureau of Investigation (FBI) agent Ali Soufan noted the group was able to grow 'its power ... by inserting itself into local civil wars and

---

83 "Selfless Six' Mourned in Kenya after Siege', *BBC News*, 22 January 2019, available at https://www.bbc.com/news/world-africa-46960387.
84 Thrall and Goepner, 'Policy Roundtable'.
85 Pam Benson, 'Bin Laden Documents: Fear of Drones', *CNN*, 3 March 2012, available at http://security.blogs.cnn.com/2012/05/03/bin-laden-documents-fear-of-drones/.
86 Ali Soufan, 'Al Qaeda Is Stronger Now Than When Bin Laden Was Killed', *The Daily Beast*, 30 January 2017, available at http://www.thedailybeast.com/articles/2017/05/07/al-qaeda-is-stronger-now-than-when-bin-laden-was-killed.

wider geopolitical conflicts'.[87] Then US Director of National Intelligence James Clapper also admitted in an interview in 2014 that US operations in the region created space for IS to expand, before concluding: 'We underestimated ISIL [i.e. the Islamic State] and overestimated the fighting capability of the Iraqi Army'.[88]

Similarly, while current military operations have allowed Iraqi troops to roll back the territorial gains of IS, they have left behind a fragmented military that still struggles to provide security for large parts of its population. Remote warfare also does not appear to be strengthening the foundations of the Afghan National Army (ANA), nor the SNA in Somalia, at a time when the territorial gains of terrorist groups on their soil are increasing. And it has left Syria an even more fractured society, with few prospects for uniting those that defeated IS with the population they now plan to control.

With this in mind, President Trump's recent decision to withdraw all US troops from Syria now that 'our coalition partners and the Syrian Democratic Forces have liberated virtually all of the territory previously held by ISIS in Syria and Iraq' seems premature (and has subsequently been delayed and undermined by members of his own administration).[89] Even as IS was confined to a tiny part of the Syrian village of Baghouz in February 2019, few commentators believed that the fight against it was over. On 25 February 2019, in a discussion with *Al-Jazeera* entitled 'Is it all over for ISIL in Syria?', analysts highlighted the importance of reconstruction, strengthening local security forces and addressing the root causes of conflict – such as bad governance and corruption – in creating truly lasting stability in the region.[90] On the same day, Rasha Al Aqeedi, of the Foreign Policy Research Institute, noted that, while IS had been pushed out of Iraq, ongoing instability, crime and violence remains 'absolutely draining [for Iraq] because … [it] does not have the ability to fight this kind of battle,

---

87 Ali Soufan, 'Al Qaeda Is Stronger Now Than When Bin Laden Was Killed '.

88 Sebastian Payne, 'Obama: U.S. Misjudged the Rise of the Islamic State, Ability of Iraqi Army', *Washington Post*, 28 September 2014, available at https://www.washingtonpost.com/world/national-security/obama-us-underestimated-the-rise-of-the-islamic-state-ability-of-iraqi-army/2014/09/28/9417ab26-4737-11e4-891d-713f052086a0_story.html.

89 Jennifer Hansler and Barbara Starr, 'Trump Touts Gains against ISIS, Glosses over Syria Withdrawal', *CNN*, 6 February 2019, available at https://www.cnn.com/2019/02/06/politics/trump-isis-conference-remarks/index.html.

90 'Is It All Over for ISIL in Syria?', *Al-Jazeera*, 25 February 2019, available at https://www.aljazeera.com/programmes/insidestory/2019/02/isil-syria-190224194712387.html.

the interior forces are just not strong enough'.[91] At the same time, the 2019 *Worldwide Threat Assessment of the US Intelligence Community* said that 'ISIS still commands thousands of fighters in Iraq and Syria, and it maintains eight branches, more than a dozen networks, and thousands of dispersed supporters around the world, despite significant leadership and territorial losses'.[92] As such, it recommended that the US should continue to engage in the region while '[t]he underlying political and economic factors that facilitated the rise of ISIS persist'.[93]

The case studies discussed in this chapter present plenty of reasons to be concerned that the use of remote warfare to disrupt terrorist safe havens may inadvertently yield exactly the kind of states – weak, ineffectual 'quasi-states'[94] – within which terrorist networks thrive, producing the very sanctuaries for terrorism the US and its allies are seeking to eliminate. It should be clear that 'with and through' does not necessarily mean 'easy, clean, and efficient' or risk-free for broader international foreign and security policy.[95]

In the end, it seems clear that terrorist activity can only be disrupted in the long-term if military operations advance the conditions for peace over the conditions for further conflict. This not only requires the military contribution to be sufficiently and consistently resourced, but requires a greater effort to address larger, institutional problems – such as fragmented, weak and even abusive domestic security sectors. A narrow focus on disrupting safe havens oversimplifies the much more complex problem of state fragility and the inadequate provision of security in many states across the world. The test for remote warfare, it seems, will be in whether it prepares the ground for greater stability or whether it locks everyone into a never-ending cycle of violence in a growing list of current and potential safe havens.

---

91 Rasha Al Aqeedi, 'With the Caliphate Crushed, What's Next?', *War on the Rocks*, 25 February 2019, available at https://warontherocks.com/2019/02/wotr-podcast-with-the-caliphate-crushed-whats-next/.

92 Daniel Coats, 'Worldwide Threat Assessment of the US Intelligence Community', Statement for the Record: Senate Select Committee on Intelligence, 29 January2019, available at https://www.dni.gov/files/ODNI/documents/2019-ATA-SFR---SSCI.pdf?utm_source=Gov%20Delivery%20Email&utm_medium=Email&utm_campaign=Media%20Contacts%20Email.

93 Coats, 'Worldwide Threat Assessment of the US Intelligence Community'.

94 Ken Menkhaus, 'Quasi-States, Nation-Building, and Terrorist Safe Havens', *Journal of Conflict Studies* 23:2 (2006), available at https://journals.lib.unb.ca/index.php/JCS/article/view/216.

95 Stein, 'Partner Operations in Syria: Lessons Learned and the Way Forward'.

References

Al-Jazeera. 'Is It All over for ISIL in Syria?', 25 February 2019.

Aslam, W. *Terrorist Relocation and the Societal Consequences of US Drone Strikes in Pakistan*. Oxford: Oxford Research Group, 2014.

Azizi, S. 'America Can't Afford to Keep Losing the War in Afghanistan', *The National Interest*, 2017.

Bandow, D. 'The Nation-Building Experiment That Failed: Time For U.S. To Leave Afghanistan', *Forbes*, 2017.

Bergen, P. 'Al Qaeda, the Organization: A Five-Year Forecast', *The ANNALS of the American Academy of Political and Social Science* 618:1, 2008.

Bergen, P. and L. Footer. 'Defeating the Attempted Global Jihadist Insurgency: Forty Steps for the Next President to Pursue against Al Qaeda, Like-Minded Groups, Unhelpful State Actors, and Radicalized Sympathizers', *The ANNALS of the American Academy of Political and Social Science* 618:1, 2008.

Carter, S.A. 'Afghanistan Quickly Becoming Safe-Haven for Terror Groups, Again', *Circa*, 15 September 2016.

Dworkin, A. *Europe's New Counter-Terror Wars*. Brussels: European Council on Foreign Relations, 2016.

Haid, H. 'The West's Obsession with ISIS Has Not Helped Syria', *CNN*, 7 April 2017.

Hoffman, B. 'A Counterterrorism Strategy for the Obama Administration', *Terrorism and Political Violence* 21:3, 2009.

Ignatius, D. 'How ISIS Spread in the Middle East', *The Atlantic*, 29 October 2015.

Jenkins, B.M. 'Disrupting Terrorist Safe Havens', RAND (blog), 18 August 2014.

Joshi, S. 'Future Wars Will Need a More Versatile Response', *The Daily Telegraph*, 13 July 2015.

Knowles, E. and A. Watson. *All Quiet On The ISIS Front: British Secret Warfare In The Information Age*. Oxford: Oxford Research Group, 2017.

Knowles, E. and A. Watson. 'How Can We Win? Lessons Learned from Contemporary Theatres', *Agile Warrior Quarterly*, 26 April 2018.

Knowles, E. and A. Watson. *No Such Thing as a Quick Fix: The Aspiration–Capabilities Gap in British Remote Warfare*. Oxford: Oxford Research Group, 2018.

Knowles, E. and A. Watson., *Remote Warfare: Lessons Learned from Contemporary Theatres*. Oxford: Oxford Research Group, 2018.

Malley, R. 'What Comes After ISIS?', *Foreign Policy* (blog), 10 July 2017.

Mansour, R. and J. Faleh. 'The Popular Mobilization Forces and Iraq's Future', Carnegie Middle East Center, 28 April 2017.

Mansour, R, and E. Van Veen. 'Iraq's Competing Security Forces After the Battle for Mosul', *War on the Rocks*, 25 August 2017.

Menkhaus, K. 'Quasi-States, Nation-Building, and Terrorist Safe Havens', *Journal of Conflict Studies* 23:2, 2006.

Payind, A. 'Inside Afghanistan 23 Years after the Soviet Withdrawal', *Journal of Asian and African Studies* 48:2, 2013.

Soufan, A. 'Al Qaeda Is Stronger Now Than When Bin Laden Was Killed', *The Daily Beast*, 30 January 2017.

Stein, A. 'Partner Operations in Syria: Lessons Learned and the Way Forward', Atlantic Council of the US, July 2017.

Turse, N. 'From Afghanistan to Somalia, Special Operations Achieve Less With More', *War Is Boring* (blog), 10 January 2018.

Verhoeven, H. 'The Self-Fulfilling Prophecy of Failed States: Somalia, State Collapse and the Global War on Terror', *Journal of Eastern African Studies* 3:3, 2009.

Zenko, M. and A.M. Wolf.,'The Myth of the Terrorist Safe Haven', *Foreign Policy* (blog), 16 January 2015.

CONCLUSION
THE FRUSTRATING
PHENOMENA
OF STATE FRAGILITY

Where to, Why to, Who to, How to ...

David Brown[1]

The problem of state weakness – in all of its various permutations, on the seeming sliding scale from fragility to absolute failure – has become a central feature of the international security environment for much of the post-Cold War era. Although its antecedents predate the events of 11 September 2001 (9/11), it was the seeming link between the action or inaction of the Taliban and the planning and execution of Al-Qaeda's simultaneous attack on multiple targets in the United States homeland that propelled such phenomena into a more central position within the security agendas of leading Western powers. This was epitomised in the George W. Bush administration's eminently challengeable assertion in 2002 that 'America is now threatened less by conquering states than we are by failing ones',[2] while analysts such as Kaplan referred to problems emanating from such states as 'a menace unlike any other'.[3] Such actors – and the grandiose rhetoric that seems to envelop them – remain a key characteristic of the 21st century security landscape for academics and practitioners alike.

Even the contemporary popular press is replete with horror stories from states seemingly teetering on the edge of what Sorenson referred to

---

1   This chapter solely reflects the views of the author and not the Royal Military Academy Sandhurst, the UK Ministry of Defence or the British Army.
2   *The National Security Strategy of the United States* (Washington DC: White House, 2002) available at https://georgewbush-whitehouse.archives.gov/nsc/nss/2002/.
3   Seth D. Kaplan, *Fixing Fragile States: A New Paradigm for Development* (London: Praeger Security International, 2008), 1.

as when 'anarchy has become domesticated'.[4] In 2019, there were sustained challenges, both internal and external, to the Maduro regime in Venezuela, presiding over corruption, hyper-inflation (estimated to be at 1,300,000 per cent at the end of 2018),[5] severe shortages of medicine and food – where it is believed that one in five go without sufficient food[6] – and a continued Western and regionally backed challenge to the legitimacy of the Maduro presidency by the leader of the National Assembly, Juan Guaidó. At the same time, the continued rending of the Libyan state, torn between three competing governments and an estimated 127 different militia groups,[7] as noted in this volume by Goher, rivals continuing chaos in Yemen.[8] Given, as Pospisil notes in his chapter here, the relatively scarce number of empirical studies detailing the nature of what is clearly a growing problem, relative to more abstract or generic assessments of the nature of fragility more generally, it is important to recall at the outset here the range of diverse geographical assessments contained in this book, from the aforementioned Libya to Somalia, to the wider context of African states seemingly bucking regional trends, such as Senegal, Uganda, Rwanda, and Ethiopia, through to Columbia and Thomas-Llewellyn's series of temporal and regional examples from Rwanda in the 1990s, via West Africa and specific responses to a wider health crisis and then finally to Myanmar.

These disparate regional case studies not only re-emphasise the international dimension of this issue, but also add to the existing complexity confronting any analyst or practitioner seeking to navigate through such murky waters. In this concluding chapter, drawing heavily on an already burgeoning literature – which, at times, seems to generate more heat than light – and the contributions in this volume, four areas of continued complexity will be focused on. Firstly, the definitional issue must be considered, with an attempt made to clarify some of the conceptual cul-de-sacs that the

---

4   Georg Sorenson, 'After the Security Dilemma: The Challenges of Insecurity in Weak States and the Dilemma of Liberal Values', *Security Dialogue* 38:3 (2007), 365.

5   'Venezuela: All you need to know about the crisis in nine charts', *BBC News*, 4 February 2019, available at https://www.bbc.co.uk/news/world-latin-america-46999668.

6   Stephen Gibbs, 'Fifth go hungry in Venezuela', *The Times*, 17 July 2019, 28.

7   Andrea Carboni and James Moody, 'Between the cracks: Actor Fragmentation and local conflict systems in the Libyan civil war', *Small Wars and Insurgencies* 29:3 (2018), 464.

8   For details on the continued problems within Yemen, see, for instance, International Crisis Group, *The Houthis: From Saada to Sanaa: Middle East Report 154*, (Brussels: International Crisis Group, 2014); Alexandra Lewis, 'Violence in Yemen: Thinking about Violence in Fragile States beyond the confines of conflict and terrorism', *Stability: International Journal of Security and Development* 2:1 (2013), 2–28; Asher Orkaby, 'Yemen's Humanitarian Nightmare: The real roots of the conflict', *Foreign Affairs* 96:6 (2017), 93–5.

existing literature has created, to lighten the ever-increasing lexicon that has developed around this phenomenon. Rather than seek to add to the proliferation of terms and models that already litter the existing secondary literature still further – 'there are as many lists of fragile states as there are definitions'[9] – the first section here will seek to re-examine some of the continued conceptual obstacles to greater understanding of the wider subject, in and of itself. This matters not least because the scale and scope of the problem will significantly shift depending on how its nature is framed, with the World Bank warning that intervenors should not 'let perfection be the enemy of progress'.[10] Of particular interest will be the importance of context, both temporal and geographical, the need to delineate more clearly between capacity and intent, the seemingly counterintuitive normative attraction of the labels of 'state failure' or 'fragility' and the stubborn continuation of the use of Western lenses – which Chandler, in his chapter, argues can never be fully or effectively overcome, leaving the analyst and practitioner trapped in 'a modernist deadlock', where all intervenors are weakened by 'their inability to see the problem in the ways in which it might appear to those more closely involved'.

The second and third sections here also relate to the seeming dichotomy between international and both regional and local actors within a specific neighbourhood. In the second section, a critical assessment is made of security challenges presented by perceived state weakness – notably the Afghan-inspired link between state failure in its many forms and the prevalence of international terrorist activity. Having critiqued some of the stated justifications for wider international interventions, the third section focuses on *who* is best placed to intervene, reappraising the seeming binary choice between regional and international actors. Finally, the last section will explore some of the potential responses proposed, both within the wider academic literature – in no sense an exhaustive list, given the array of alternative approaches contained within – and advanced by contributors to this volume, from the different forms and scales of military intervention, including more limited applications, such as the use of drones and the role of Security Sector Reform (SSR) to train and equip home-grown forces, as well as larger scale military campaigns in states such as Afghanistan, which has incorporated limited approaches within a more extensive campaign

---

9 Kaplan, *Fixing Fragile States*, 6.
10 World Bank cited Roger MacGinty, 'Against Stabilization', *Stability: International Journal of Security and Development* 1:1 (2012), 21.

that has lasted longer than the duration of the First and Second World Wars combined. 'Softer' methods, such as the use of international aid and the promotion of power-sharing peace processes, as explored by Pospisil in his chapter, will also be considered. While it may be that the problem – in terms of its scope and severity in international security terms – is less than the early rhetoric had suggested, equally the international community (both Western and regional actors) has yet to locate a suitable combination of commitment and contribution that would sufficiently strengthen the resilience of individual target states, to remove what Rodriguez and Sanchez in their chapter here refer to as 'the failed state stigma', a somewhat provocative label given the 'naming and shaming' attitude attached to such an approach.

The Definitional Dilemma: An Inescapable Trap?

While there is no accepted and agreed definition of the wider concept of state fragility, this is not for want of effort, as academics and key agencies within national and international governance, such as the World Bank and the Organisation for Economic Cooperation and Development (OECD) have sought definitions to suit their theoretical or specific functional interests.[11] The literature teems with terminology, revolving around shared concepts and concerns across the spectrum of governance, from socio-economic indicators, such as gross domestic product (GDP) and annual growth rates, to infant and maternal mortality, to political concerns, such as the seemingly ubiquitous nature of official corruption at multiple levels of authority and the emergence of alternative extremist ideologies refusing to accept the existing legitimacy of governance or underlying state structures, leading to what Goher calls a level of 'competition that renders the notion of loyalty to the state meaningless for almost all … parties'.

---

11 For a flavour of the wider literature in this area, see Lothar Brock et al, *Fragile States: Violence and the Failure of Intervention* (Cambridge: Polity, 2012); Maria Gabrielsen Jumbert, 'How Sudan's 'Rogue' label shaped US responses to the Darfur crisis: What's the problem and who's in charge?', *Third World Quarterly* 35:2 (2014), 284–99; Sonja Grimm, Nicolas Lemay-Hebert and Olivier Nay (eds), *The Political Invention of Fragile States: The Power of an Idea* (London: Routledge, 2015); Jennifer Miller and Keith Krause, 'State Failure, State Collapse and State Reconstruction: Concepts, Lessons and Strategies', *Development and Change* 33:5 (2002), 753–74; Mohammed Nuruzzaman, 'Revisiting the category of Fragile and Failed States in International Relations', *International Studies* 46:3 (2009), 271–94; Abel Polese and Ruth Hanau Santini, 'Limited Statehood and its Security Implications on the fragmentation of political order in the Middle East and North Africa', *Small Wars and Insurgencies* 29:3 (2018), 379–90; Robert I. Rotberg (ed.), *When States Fail: Causes and Consequences* (Princeton NJ: Princeton University Press, 2004).

At one level, the literature focuses on the more technocratic – and allegedly politically neutral – aspect of state-building, emphasising the need to build new capacities upon existing structures and territories, most notably in the field of legislative and law enforcement, but also across bureaucracy as a whole. At another, and setting an even more difficult task, particularly for external intervenors, parts of the literature advocate the fundamental importance of nation-building, in terms of creating a more cohesive identity above the level of the clan, tribe, village or region, despite the claim of some, such as Mason, that 'no nation has ever been built' externally,[12] even with an increased emphasis on education, culture and, perversely, the solidification of internal identities in opposition to their continued external presence. As part of an occasional academic tendency to decant older wine into shiny newly labelled bottles, the overarching phenomenon has been described, at different times, as 'failed', 'failing', 'fragile', 'weak', a 'quasi-state', a 'collapsed' state, an 'anarchic' state and a 'phantom' state,[13] to name but a few. These concepts either draw directly from, or explicitly reject, the Weberian model of sovereign statehood, with its focus on levels of control through an identifiable and legitimate monopoly over the use of force within a given territory. Explorations have been made as to the relationship between capacity, effectiveness and wider legitimacy,[14] taking into account a multiplicity of internal and external audiences, who may not share the same normative value bases and therefore could respond differently to the application of capacity in different policy areas, regional dimensions and over time. One classic example from the State Failure Project, for instance, made explicit reference to the 'failed state' being 'utterly incapable of sustaining itself as a member of the international community',[15] which seems to subtly shift the focus of attention from the

---

12 M. Chris Mason, 'Nation-Building is an oxymoron', *Parameters* 46:1 (2016), 68.

13 In addition to the references in note 11, see Jennifer G. Cooke and Richard Downie, *Rethinking Engagement in Fragile States* (Washington DC: Center for Strategic and International Studies, 2015), available at https://csis-prod.s3.amazonaws.com/s3fs-public/legacy_files/files/publication/150722_Cooke_RethinkingEngagement_Web.pdf; Helge Mari Loubser and Hussein Solomon, 'Responding to state failure in Somalia', *African Review* 6:1 (2014), 1-2; Robert I. Rotberg, 'The New Nature of Nation-State Failure', *Washington Quarterly* 25:3 (2002), 85–96.

14 See, for instance, Jack Goldstone, 'Pathways to State Failure', in Harvey Starr (ed.), *Dealing with Failed States: Crossing Analytical Boundaries* (London: Routledge, 2009); Kaplan, *Fixing Fragile States*.

15 Michael Stohl and George Lopez, *Westphalia, the End of the Cold War and the New World Order: Old Roots to a "NEW" Problem'*. Paper delivered at the 'Failed States and International Security: Causes, Prospects, and Consequences' conference in February 1998, (West Lafayette: Purdue University, 1998), available at http://mstohl.faculty.comm.ucsb.edu/failed_states/1998/papers/stohl-lopez.html.

state itself – in terms of how effectively central government particularly develops and discharges its capacities in relation to maintaining and improving the security, health and well-being of its people – to how the state behaves and reacts internationally, with such looser terminology opening up an array of potential benchmarks, both in terms of how the state protects its borders but also the manner in which it conducts international diplomacy. In effect, for all the different permutations and combinations of differing diagnoses that exist at present, potentially bewildering for both academic and practitioner alike, as Menkhaus has contended, there is actually a 'broad consensus on the general traits of state fragility and failure' but 'not on the specifics of how to measure them and weigh them'.[16] The constituent concerns are relatively similar, but the inter-relationship between them, the relative emphasis given to them by differing actors, both local, regional and international, and the impact and implications of them, will differ on a case by case basis.

Such differences on the details is hardly surprising given the importance of context in all its many and varied forms. As Goldstone noted, paraphrasing Tolstoy's opening to *Anna Karenina*, 'all stable actors resemble one another, each unstable nation is unstable in its own way'.[17] While there are identifiable common traits across the case studies in this volume, as well as within the wider academic literature, there is no formula for failure that can or ever will be defined that would apply in each and every given set of circumstances. This is an underlying weakness of universal lists that seek to rank and rate *all* states against an agreed list of criteria, including arguably the most prominent, the Fund for Peace's annual Fragile State Index. As Innes and Booher have noted, even if you could identify and isolate the primary general causes of state weakness – without taking into account what would provoke further deterioration in the state's overall effectiveness, shifting it further down the superficially attractive linear relationship between fragility and absolute failure, where 'anarchy becomes more and more the norm'[18] – 'causality cannot be definitively established and, because the system is constantly subject to unanticipated change, the idea of a best solution is a mirage'.[19] In effect, as with the concept of power more generally – itself a continually contested concept among

16 Kenneth J. Menkhaus, 'State Fragility as a Wicked Problem', *Prism* 1:2 (2010), 88.
17 Jack Goldstone et al cited Menkhaus, 'State Fragility as a Wicked Problem', 89.
18 Rotberg, 'The New Nature of Nation-State Failure', 87.
19 Innes and Booher cited Cedric de Coning, 'Adaptive Peacebuilding', *International Affairs* 94:2 (2018), 313.

scholars – there is a need to better identify and measure the effectiveness of governmental capacity, as much as is possible, over time and in different geographical contexts, both within a state and between states. In the case of the latter, if a state's effective reach in terms of the projection of central capacity differs from region to region, as has been the case in numerous examples, what would be the most appropriate label to use?

Rotberg has noted that 'plausibly the extent of a state's failure can be measured by the extent of its geographical expanse genuinely controlled by the official government'.[20] Leaving some of his caveats to one side for the moment – both in terms of the 'genuine' nature and 'official government' – Rodríguez and Sánchez take such quantitative measurements further in their chapter here, arguing that the effectiveness of the Colombian state depends not only on simple geographical reach, but also on the significance of the regions where the state's writ is more or less consequential. In the case of the former, the duration of a state's seeming ineffectiveness – while complicating the practicalities of external intervention, with early signs of state fragility being both easier to resolve (at least in theory) and more difficult to muster the political necessity for intervening in the first place – has been a central feature of the wider debates, with Rotberg again emphasising the enduring nature of underlying weaknesses, especially in terms of the extent and intensity of violence.[21] The OECD has shifted from an overarching view of a 'fragile state' to a focus on 'states of fragility', seeking to separate out long-term fragility across governance as a whole to more fleeting, policy-specific moments.[22] As a consequence, it is worth remembering Weiss's warning from over 20 years ago that 'the idea of a generalised state capacity is meaningless',[23] both because of the disparate nature of its impact, and because its effectiveness needs to be tested, to see how the state responds to both unexpected and structural crises. On paper, the state may seem strong and effective, because it is neither trying nor being asked to do too much. As a consequence, the seeming depth of existing capacity can be beguiling, but, as with the phenomenon of so-called 'ghost soldiers', who appear on lists but not in battle, apparent quantitative assets in and of themselves are not a sufficient test of the effectiveness of a state's capacity.

Not only that, but – given the more nuanced picture that emerges when the intervening factors of time and space, as well as wider

---

20 Rotberg, *When States Fail*, 6.
21 Rotberg, 'The New Nature of Nation-State Failure', 85.
22 Cooke and Downie, *Rethinking Engagement in Fragile States*, 2–3.
23 Linda Weiss, *The Myth of the Powerless State* (New York: Cornell University Press, 1998), 4.

contextual appreciation, are taken into account – there needs to be a wider acknowledgement that the usefulness of 'failed' state as an accurate label has significantly waned, to the point where analytical attention and practitioner focus should be addressed to the early symptoms, and not to what Fisher describes in his chapter as 'the end point, at which political, social, economic and security institutions disintegrate, sometimes simultaneously'. In seeking to lighten the lexicon – which often somewhat promiscuously uses the terms 'fragility', 'failing' and 'failed' interchangeably, as if they were not differing stages within a wider process, that neither has to be linear nor even – it is best to acknowledge the admission of the compilers of the aforementioned Fragile States Index, that, having deliberately used the term *Failed* States Index at the outset in 2005 even though it was 'fraught with issues' given its more absolute status, 'in order to get attention', they chose to shift to the more accurate Fragile States Index, as 'we found that we were having more conversations about terminology than substance'.[24] Take the case of Somalia – along with the Democratic Republic of Congo (DRC) and Afghanistan, one of the consistent 'poster boys' of the Fragile States Index, joined in more recent times by South Sudan, whose oil potential has been effectively undermined by poor governance, external conflict with Sudan, internal conflict between competing ethnic and political groupings and extensive patronage and corruption practices.[25] Although Somalia was considered by Rotberg to be beyond a failed state – a 'collapsed' state, in his view[26] – this image, although consistently held to in wider discourse, fails to capture both the progress that has been made over time[27] and the existence of both Puntland and Somaliland, neither of which are externally recognised states, but have demonstrated significant local capacity to govern. Menkhaus, in describing Somalia as 'a loose constellation of commercial city states and villages, separated by long stretches of pastoral statelessness', argues that, even as central government seemed to collapse, there was 'an impressive if fragile level' of local governance that survived and, in the aforementioned cases of Somaliland and Puntland, seemed to

---

24 Krista Hendry, 'From Failed to Fragile: Renaming the Index', in *Fragile State Index 2014* (Washington DC: Fund for Peace, 2014), 1.

25 For a personal reflection on and academic analysis of the development of the relatively new state of South Sudan, see Hilde F. Johnson, *South Sudan: The Untold Story. From Independence to Civil War* (London: IB Tauris, 2016).

26 Rotberg, 'The New Nature of Nation-State Failure', 90.

27 See, for instance, Bronwyn Bruton, 'In the Quicksands of Somalia: Where doing less helps', *Foreign Affairs* 88:6 (2009), 79-94; Stephen Burgess, 'A Lost Cause Recouped: Peace Enforcement and State Building in Somalia', *Contemporary Security Policy* 34:2 (2013), 302–23.

thrive.[28] There is thus a question to be asked regarding how much account should be taken of *alternative* sources of governance, both public and private, in terms of service delivery and general governance, in determining the nature of the problem and advocating and allocating resources in response. As Berdal has argued, 'the absence of government does not invariably mean the absence of governance',[29] particularly given that, for many of its citizens, contact with central government will have been and may remain relatively limited, as was the experience in Afghanistan.[30] The centrality of the centre should not hold in assessing state fragility to the extent it has in the past – to where the concept of 'failed state' should ultimately be consigned.

Greater awareness of the context within which the state is seeking to apply its existing capacity – geographically, temporally and in relation to specific crises and events – has to be further tempered by Chandler's prescient but pessimistic warnings in his chapter here regarding the continued prevalence of Western biases, both in terms of the benchmarks that are set to measure the extent of state failure but, as importantly, in terms of the level of knowledge accessible to Western scholars and practitioners, when both identifying problems and formulating responses. In effect, he argues that, beyond Ikenberry's description of the problem of state fragility as one where the investigated become 'prisoners of a fixed genetic code',[31] the investigators are similarly trapped by a pre-determined set of beliefs that bring with them loaded assumptions and pre-existing limitations that cannot be dealt with by increased cultural immersion or a greater faith in the development of new technology to expand ontological horizons to a level where a sufficient basis of appreciation would exist, even to assist with a bottom-up approach to state-building. It is not clear if and how Thomas-Llewellyn's advocacy of greater involvement from within fragile states themselves – in the construction of 'an agreed operational framework' that would, at the very least, seek to 'identify the range of disciplines' necessary to formulate a more coherent appreciation of what the military as an actor would need to understand individual cases of state fragility – would

---

28 Kenneth J. Menkhaus, 'Vicious Circles and the Security Development nexus in Somalia', *Conflict, Security and Development* 4:2 (2004), 153–60.

29 Mats Berdal, *Building Peace after War* (London: Routledge, 2009) 123.

30 Barakat and Larson have argued in relation to Afghanistan that a focus on the executive level misses 'critical dynamics at the regional or sub state level'. Sultan Barakat and Anna Larson, 'Fragile States: A Donor Serving Concept? Issues with Interpretation of Fragile Statehood in Afghanistan', *Journal of Intervention and Statebuilding* 8:1 (2014), 22.

31 G. John Ikenberry cited Weiss, *The Myth of the Powerless State*, 196.

ameliorate Chandler's concern that even 'the understanding that there is "a problem" constitutes a fundamental gap between the statebuilding agency' in whatever form 'and the society concerned'. How to both appreciate and ameliorate such conceptual limitations and discrepancies should therefore provide a productive avenue for continued academic research and debate in the future.

In seeking to inject an element of practical pragmatism into an essentially theoretical discussion regarding the self-imposed limits of Western knowledge creation, Thomas-Llewellyn here notes that, while being aware of conceptual lenses and their limitations, there is a practical danger of them becoming blinkers, leading us on the road to nowhere. Acknowledging the admonition that the best should not be the enemy of the good – although it is debatable even how close in certain cases external intervenors have come to that more limited objective – there is a need to avoid two scenarios in which both the intervened and intervenor are effectively set up to fail. In the case of the former, by both over-emphasising and under-delivering on democratic mechanisms as a means to stabilise a given situation, whilst the wider international community is in danger of setting standards the state in question – and arguably its wider society – is not ready to undertake within a pre-determined timetable. Pospisil's findings from his extended data set on peace processes are useful here. He concludes that while 'the essential features of a functional democratic polity' were a regular feature of negotiated agreements, significantly, the 'wider software' necessary to underpin a 'strong and stable' democracy, such as support for civil society and political party reform, were just as regularly either missing or 'at best superficially addressed'. Both demonising and making a dash for democracy only underlines Chandler's concern that external stabilisers are primarily focused on 'surface appearances', rather than underlying and more complex root causes.

However, that should not necessarily lead the West in particular to dismiss the value of democracy as a potentially stabilising force. There are already wider concerns about the future stability of the liberal international order, whether as part of some Trumpian US retrenchment or as a normative consequence of the nascent multipolar world; although beyond the scope of this volume, it would be an interesting analytical avenue to explore further how rising powers, such as Russia, China, India and Brazil, seek to identify and potentially challenge pre-conceived Western notions of what constitutes state fragility. As Halverson argues, 'it is not the empirical character of these authorities that has potentially changed, but the frameworks through which they are evaluated'.[32] If failure and fragility are, to a large extent, in the eyes of the beholder, then the more eyes on the problem, the more difficult it will become to reach even a working consensus. Chandler's concerns, while conceptually compelling, do not ultimately aid the practitioner in seeking a resolution to the problem. Awareness of conceptual difficulties should not lead to the abandonment of the wider cause of seeking to better understand and respond to the existing phenomena of state fragility; instead, taking his concerns fully on board, there is a need to seek as much as possible to mitigate such limitations, whilst accepting that they will never be fully mastered.

It should also be noted that the general assumption that state failure is not to be welcomed is not shared by all. There needs to be an acknowledgement of the practical *advantages* (for some) of failure, in effect that, counterintuitively, there is power in weakness. Matisek is in good company when he notes that the prevalence of perceived weakness across Africa – 'Africa is full of weak states' – may not solely be as a result of the ineffectiveness of existing state structures and capacity, but may also be a *deliberate* choice on the part of individual regimes, either to play on the external sympathies of potential aid-givers or as a means of positively affecting the internal balance of power in their favour, thereby seeking to protect the regime at the expense of the state. In similar vein, Goher argues that weakness was historically the norm in Libya, although a new form and depth of weakness has certainly emerged post-2011. Yemen is an even more explicit case, with Philips elsewhere claiming that 'crisis has kept the system running and has been, to a significant degree, a deliberate choice by Yemen's power elite'. Arguing that the Salah regime in particular 'mortgaged its future on its ability to bargain for external support', a pattern of behaviour is highlighted, with known jihadists and suspected Al-Qaeda operatives released from Yemeni jails on more than one occasion to create conditions of 'managed' chaos and maintain a level of financial attention from the US particularly.[33]

Finally, the issue of deliberate action leads to arguably the most significant piece of conceptual clearance work that needs to take place – namely the level of erroneous emphasis given to either the deliberate oppressive application of state capacity or equally the decision to withhold

32 Dan Halverson, *States of Disorder: Understanding State Failure and Intervention in the Periphery* (Farnham: Ashgate, 2013), 17.
33 Sarah Philips, 'Yemen and the Politics of Permanent Crisis', *Adelphi Papers* 51:420 (2011), 12–14.

it (perhaps for reasons noted above). In Rotberg's classic definition of state failure, equal weight is given as follows: 'a polity that is no longer able *or willing* to perform the fundamental tasks of a nation-state' (emphasis added).[34] While the end result may prove to be the same, there is a difference that is important to note in seeking a clearer focus and a greater level of conceptual precision. In effect, the failings of a state – whether it be extensive human rights abuses or levels of famine, the spread of disease or the persecution of minorities within society – need not automatically be assumed to be the sign of a state failing, in terms of the application and effectiveness of its capacity. The continued existence of the contested but enduring 'rogue state' phenomenon suggests there is a need to more clearly delineate between the *inability* of a state to respond and the *unwillingness* of a state to do so, or indeed to be the active source of the problem itself. By muddying the definitional waters unnecessarily, denying the state agency when such agency exists, there are implications in terms of appropriate response, between sanction and support: 'if the state in question is unable rather than unwilling ... then there is nothing to be gained by condemning it'.[35] If the state is guilty directly, rather than passively, then building up further the capacity of a state already capable and culpable seems to make little sense. The blurred lines between what constitutes 'rogue' behaviour and the signs of a state lacking sufficient capacity to respond have further implications when considering the second area of interest in this chapter, namely the nature of the threat posed by failing and fragile states.

The Afghan Analogy: Terrorism and the Nature of the International Commitment

In assessing the reasons for the limited international intervention in Libya by NATO, under the auspices of UN Resolutions 1970 and 1973, Goher notes that the West was motivated by a combination of 'value oriented moral determinants and self-interested material determinants', a blend, in effect, of liberalism and realism. In the case of the former, the humanitarian implications of a failure on the part of the state to act to resolve crises, to demonstrate sufficient resilience and to improve overall levels of human

---

34 Rotberg cited Nuruzzaman, 'Revisiting the category of Fragile and Failed States in International Relations', 273.

35 Adrian Gallagher, 'The promise of Pillar II: Analysing International Assistance under the Responsibility to Protect', *International Affairs* 91:6 (2015), 1264.

security form part of the perceived paradigm shift from a 'right to intervene' on humanitarian grounds to the much-vaunted Responsibility to Protect (R2P), which bases its applicability both on states unwilling and unable to act in the face of developing humanitarian crises. In the case of the latter, as well as potentially threatening the credibility of the wider international community in terms of upholding certain standards of behaviour – what Halverson refers to as 'challenging their role identity as the source and guarantor of international order'[36] – there is also the perceived link between a lack of sufficient state capacity and the increased room for manoeuvre of international and domestic terrorist groups. In reality, rather than be misled by the rhetoric that has emerged in both academic and political circles, there are continued and increasing limits on the willingness of international actors particularly to sustain a humanitarian intervention, with R2P structurally undermined in such a way that it has not lived up to its potential.[37] Additionally, the causal relationship between state fragility and the incidence of terrorist activity is less clear-cut than some proponents initially suggested.

In part this relates to a misreading of the nature of the relationship between Al-Qaeda under Osama bin Laden in the late 1990s and the Taliban government in Afghanistan under Mullah Omar. Rather than justify the development of the terrorist training infrastructure that emerges in Afghanistan in this period as a consequence of even benign neglect, the relationship is far more activist, even if it has been claimed that the Taliban were unaware of the details of the specific planning that was undertaken prior to 9/11.[38] It was not a case that the Taliban were unaware of the existence of training camps, nor the wider intentions of their 'honoured guests', with Crumpton noting that, 'Al Qaeda, if anything, had co-opted the Taliban leadership', while 'still taking advantage of their stunning ignorance of world affairs'.[39] In that sense, by effectively turning a blind eye, at best, or actively supporting, at worst, the state-terrorist nexus is not one

---

36 Halverson, *States of Disorder*, 4.
37 For a flavour of the wider literature on this issue, see Alex J. Bellamy, 'The Responsibility to Protect: Added Value or Hot Air?', *Co-operation and Conflict* 48:3 (2013), 333–57; Simon Chesterman, 'Leading from Behind: The R2P, the Obama Doctrine and Humanitarian Intervention after Libya', *Ethics and International Affairs* 25:3 (2011), 279–85; Gareth Evans and Ramesh Thakur, 'Humanitarian Intervention and the Responsibility to Protect', *International Security* 37:4 (2013), 199–214; Ramesh Thakur, 'The Responsibility to Protect at 15', *International Affairs* 92:2 (2016), 415–34.
38 Steve Coll, *Directorate S: The CIA and America's secret wars in Afghanistan and Pakistan 2001–2016* (London: Allen Lane, 2018), 68.
39 Hank Crumpton cited Coll, *Directorate S*, 69.

based on a lack of sufficient capacity, in terms of oversight or enforcement, but is closer in nature to the traditional activist idea of state-sponsorship of terrorism, that seems fortuitously to have dwindled in importance within the wider international security environment in recent times.[40] NATO's justification in part for invoking Article 5 in the wake of 9/11 was based on acknowledging the nature of the relationship between the Taliban and Al-Qaeda, in terms of state level involvement, even if that relationship was not as linear as the traditional state-sponsorship model most regularly associated with – although equally disputed by – Iran would suggest.

Nor is the Afghanistan analogy an isolated example, even as it is misapplied in the wider literature: both Hehir and George have systematically sought to deconstruct the relationship between state fragility and differing forms of terrorism and have not found a direct causal link of note, certainly not with transnational or international terrorism. In fact, their statistical evidence suggests a more nuanced picture, with only three of the top 20 states ranked in the Fragile States Index believed to act as 'breeding grounds' for terrorist activity, while the majority of that group – 13 out of 20 – showed little or no evidence of significant terrorist activity at all.[41] Of these, two of the three noted cases at the time – which is over a decade ago now – were Iraq and Afghanistan, which may have become more prevalent grounds for terrorist and wider insurgent behaviour to a large extent because of Western intervention not simply undermining but deconstructing the capacity of the pre-existing state to act. In Knowles and Watson's study of security responses to so-called 'safe havens' for international terrorism in this volume, they label the West 'the bringers of failure' in this context. Elsewhere, George has come to the conclusion that such states are 'attractive mostly for [terrorist] perpetrators of the same country and to an extent neighbouring states' – an important distinction to bear in mind when considering the appropriate level of response below – rather than as a base for international terrorists to act: 'it is highly unlikely that foreign perpetrators use failed states as safe havens, because of the

---

40 See, for example, Daniel Byman, *The Changing Nature of State Sponsorship of Terrorism: Analysis Paper 16* (Washington DC: Brookings Institution, 2008). The US State Department lists just four state sponsors within the current international security environment: Iran, North Korea, Sudan and Syria. Bureau of Counter-Terrorism and Countering Violent Extremism, *Terrorist Designations and State Sponsors of Terrorism* (Washington DC: State Department, 2019), available at https://www.state.gov/state-sponsors-of-terrorism/.

41 Aidan Hehir, 'The Myth of the Failed State and the War on Terror: A Challenge to Conventional Wisdom', *Journal of Intervention and Statebuilding* 1:3 (2007), 315.

existing hostile conditions within these countries'.[42] Whether it be because of the lack of sufficient logistics, infrastructure and technology to assist with the planning and preparation of terrorist activity, or as a consequence of the weakened sovereign control from central government inherent in such states, 'failed states are where the terrorists are most vulnerable to covert actions, commando raids, surprise attacks'.[43] The wider evidential base is therefore not substantial enough to support a linkage; instead, it is more likely the case, as Chandler has argued elsewhere, that 'the idea of failed states as a security threat' – at least internationally – 'is an exaggerated one', based on the false analogy of Afghanistan.[44]

It is sadly not the case that a reliance on the moral dimension of Goher's intervention equation will compensate for this more inconsistent security relationship. While the rhetoric surrounding the promotion of R2P was about increasing the consistency and commitment of international responses, whether preventative, through capacity building – as part of the relatively under-utilised Pillar Two – or, in the final analysis, military in nature, the reality is that, having had little choice but to graft such commitments onto the flawed structure of the UN Security Council (UNSC), such hopes were unlikely to ever be fully or effectively realised. The rhetoric may be both defiant and definitive – take, for example, the assertion from former UK Prime Minister Tony Blair that 'if you allow a series of failed states to rise, then sooner or later you end up having to deal with them'[45] – but the reality is that inconsistency is a fact of international life, as states constantly have to make choices, even when confronted with the horrifying humanitarian consequences of both weak state capacity and external military interventions, as in the case of Yemen. Rather than presume a new era of international humanitarian endeavour, to rescue peoples from either the incapacity or intentional cruelty of their own governments, the evidence of any sea change has been sadly lacking.[46] The 2011 NATO-led intervention in Libya was heralded as a new high point in humanitarian intervention, but,

---

42 Justin George, 'State Failure and Transnational Terrorism: An Empirical Analysis', *Journal of Conflict Resolution* 62:3 (2018), 481.

43 Gary Dempsey cited Justin Logan and Christopher Preble, 'Failed States and Flawed Logic: The case against a standing Nation-Building Office', *Policy Analysis* 560 (2006), 6.

44 David Chandler cited Hehir, 'The Myth of the Failed State and the War on Terror', 308.

45 Tony Blair cited Hans Henrik Holm, 'Failing Failed States: Who forgets the forgotten?', *Security Dialogue* 33:4 (2002), 457.

46 See, for example, David Brown, 'Kosovo and Libya: Lessons Learned for Limited Humanitarianism?' and Aidan Hehir, 'Kosovo 1999: The False Dawn of Humanitarian Intervention', *Comparative Strategy* 38:5 (2019).

at most, that is a relative judgement on what had gone before; in reality, both in conception, delivery and, most importantly, in terms of the willingness of the international community generally and the West specifically to remain in-theatre and commit to a credible rebuilding of Libyan capacity, the commitment to such humanitarian endeavours remains very limited, at best.

Local Solutions to Local Problems? The Role of Regional Actors

In justifying a sufficient level of support for UN-endorsed action in Libya, it was claimed the open diplomatic and then operational support from regional actors, both states, such as the United Arab Emirates and Qatar, and regional organisations, such as the African Union (AU) and Arab League, was crucial, both in legitimising international action and warding off the international 'spoilers' of Moscow and Beijing. The multifaceted relationship between regional and international actors in tackling state fragility has been a key theme in the wider literature and is the particular focus of Fisher's chapter in this volume. Given the flaws noted, both here and more widely, in the level of commitment, awareness and effectiveness of particularly Western reactions in recent years, the need to advocate alternative arrangements, on the face of it, increases in relative importance. Regional actors are believed to be relatively better placed, in terms of the appreciation of local culture and custom, are more likely to be directly or indirectly affected given geographic proximity, and assumedly offer an alternative approach to that of the West, given they are unlikely to be as infected by the Western-centric conceptual blinkers discussed by Chandler.

However, rather than simply relying on 'inversing the focus', as Chandler notes, by advocating an almost binary switch between international and regional responses, there needs to be a wider recognition that, not only do such actors bring with them a different set of limitations, and a worrying number of all-too-familiar concerns, but, as Fisher has pointed out in relation to AMISOM's involvement in Somalia, their 'solutions' may not be markedly different to those of the West either. In fact, as Fisher notes, the experience in Somalia 'problematises the assumption that geographic proximity equates to an appreciation or respect for local forms of political authority'. In a telling tale of equally vested interests and not-too-hidden agendas, he outlines how states such as Kenya and Uganda, in as much as they are willing to reconstitute the Somali state in the first place, are prepared to do so in a way that both maintains internal fragmentation and extends the level of pre-existing weakness, so as not to endanger the regional balance of power. The thinking behind this, as has been noted elsewhere, is that 'the confirmation of a fragmented non-functional Somali state may, from the perspective of Ethiopia and Kenya, be the least bad and most manageable option'.[47] Not only that but, echoing complaints made regarding international diplomatic efforts at the 2012 London Conference on Somalia's future – where it was claimed the Somali Constituent Assembly was confronted in an internationally agreed, truncated timetable of a week to approve a new draft constitution, with non-approval not an option[48] – the engagement of regional actors has not led to a noted increase in the level of consultation or respect for local viewpoints. Fisher notes that Somalia's neighbours effectively exploited its formal label as a 'failed state' to limit consultation and propose their own models of governance, regardless of whether they would apply within a Somali context: '"Failed" status means not having to ask what locals think … Somalis are still patients in an intensive care unit – you cannot expect them to know about governance yet'. Kenya remains primarily concerned with protecting the shared border, so as to avoid any significant overspill of violence and instability into its own territory. This is not markedly different to the Trumpian view that limited counterterrorist interventions in Libya are primarily 'to put America first', rather than as part of a longer term and more comprehensive package of measures to stabilise the Libyan state per se.[49] As such, it seems that regional state reactions are generated less by solidarity and shared cultural affinities and more by a seemingly universal level of self-interest, even to the point of keeping the target state relatively unstable so as to weaken a potential rival.

Rather than a clear choice between regional and wider international actors, the likelihood is that there will be some combined involvement of both, whether in terms of formal and informal legitimisation, as noted in the Libya example, or, more practically, in terms of financing, as with the institutional relationship between the UN and AU, where a 0.2 per cent

---

47 Sally Healy, 'Rebuilding State Fragility: Somalia's Neighbours: a help or a hindrance?', in Philip Lewis and Helen Wallace (eds), *Rethinking State Fragility* (London: British Academy, 2015), 74, available at https://www.thebritishacademy.ac.uk/sites/default/files/conflict-stability-rethinking-state-fragility.pdf.

48 Laura Hammond, 'Somalia Rising: Things are starting to change for the world's longest failed state', *Journal of East African Studies* 7:1 (2013), 185.

49 Karim Mezran and Elissa Miller, *Libya: From Intervention to Proxy War* (Washington DC: Atlantic Council, 2017), 6.

levy on imports by African states is supposed to cover or reimburse the UN for up to 25 per cent of the costs of AU missions authorised by the UNSC.[50] In fact, this inter-relationship even extends to the initial motivation for greater regional engagement in the first place, as Fisher points out that 'African solutions to Africa's problems' was 'a concept first developed by US officials as a means to justify [American] withdrawal'. Rather than see it as a potential force multiplier, the international community more widely may see it as another version of Donald Rumsfeld's 'scale up to scale down',[51] a zero-sum approach that substitutes local forces for international ones and effectively sees 'burden sharing ... devolve into burden shifting, as the world invests unprepared regional bodies with unrealistic expectations'.[52] In effect, without the same resources to deploy, regional actors nevertheless seem to follow the same path-dependency as their international counterparts, prioritising their own interests, seeking to impose their own solutions, minimising their own commitment. As Chandler points out, all interventions are external, regardless of where they come from; as such, they are less of a break with the past than might have been optimistically hoped.

A Persistent Problem with Too Many Answers?

A fundamental question confronting whoever ends up leading or financing a putative state 'rescue' mission is whether to reconstruct or deconstruct, to build up or rebuild over. In undertaking the process of state-building, external intervenors approach the task laden with their own assumptions, both regarding the validity of the centralised Weberian state as an appropriate model and, as noted earlier, the type of state that needs to be created in order to provide a greater guarantee of stability, as they see it, over the medium to long term. In effect, they approach such tasks with 'a systematic bias in favour of states and sovereignty'.[53] Such a bias, understandable and inescapable, needs to at least be acknowledged, not least because of the subconscious limitations it places on the process of state-building in the first

---

50 Arthur Boutellis and Lesley Connolly, *The State of UN Peace Operations Reform: An Implementation Scorecard* (New York: International Peace Institute, 2016), 12.

51 See 'Rumsfeld's Memo of Options for Iraq War', *New York Times*, 3 December 2006.

52 Stewart Patrick, 'The Unruled world: The case for good enough governance', *Foreign Affairs* 93:1 (2014), 70.

53 Burgess, 'A Lost Cause Recouped', 316.

place. For example, by placing such emphasis on 'the notion of control', a central element of the Weberian state, MacGinty worries that intervenors eliminate 'notions that promote emancipation, autonomy and dissent'.[54] Not only that, but the affirmation bias of *state*-building, reinforcing existing capacity at the centre, closes off alternative options, such as reformulating the territorial boundaries of the state, and prioritises the centre at the expense of other elements of governance within a given state.

In the case of territorial affirmation, with states seeking to primarily work within the given borders recognised internationally, even if they are part of the sources of tension internally and undermine both national cohesion and central control, Ottaway and Mair suggest that 'while the redrawing of borders is not a panacea, it should not be automatically ruled out as a means of rebuilding weak states'.[55] Such critical approaches, while performing a useful theoretical service, need to be more realistic in their recommendations. While a critique of the Weberian state model is of conceptual value, when confronted by increased violence as a consequence of the lack of state authority and legitimacy, it is difficult to argue for an approach that unleashes the 'forces of freedom' within a given society. Letting a hundred flowers bloom may sound good in principle, but it is hard to see how this aids the intervenor in the immediacy of the moment, not least when considering the criticism that Western powers have rightly faced with regard to the impact and longer term implications of state deconstruction in Iraq. Such criticisms would also be more credible if alternative practical approaches were advocated; critiquing control and promoting freedom is not a plan that would survive first contact with the situation on the ground. It is worth remembering, as a central feature of Fisher's and Matisek's work on Africa in this volume, that, rather than assume centralised state formation as a Western fixation, regional actors in Africa have sought the centralising embrace of the state themselves. Not only that but actors as normatively distinct as Islamic State (IS) still have a centralised state-building process as at least the first step towards the restitution of the true Islamic caliphate. Until such time as a credible alternative is consistently advocated, it seems unlikely that any part of the international community will accept that 'some states should be allowed to fail' even if it is 'clear that the survival of the state comes at the expense of the survival of its citizens'.[56] In Venezuela,

---

54 MacGinty, 'Against Stabilization', 20.
55 Marina Ottaway and Stefan Mair, *States at Risk and Failed States* (Washington DC: Carnegie Endowment, 2004), 7.
56 Ottaway and Mair, *States at Risk and Failed States*, 8.

for example, it is the Maduro regime that is seen as the problem, not the state itself. Similarly, while the legitimacy of borders and boundaries, both within and between states, may remain a constant source of potential conflict, the implications of seeking revisions, to create more effective nation-states along clearer ethnic grounds for instance, not only runs contrary to international orthodoxy regarding multiculturalism, but also – as seen with IS – is hardly a guaranteed recipe for increasing stability.

In a more limited intervention – relatively speaking – there is greater concern within the academic literature about the balance struck between both the centre and the periphery within the state in question and between the reconstituted central authority and its seemingly helpful external intervenor. As part of a wider concern regarding how best to harness local knowledge, practice and governance – even with the warnings here from both Chandler regarding the limitations of a more explicit 'bottom-up' approach, which is still initiated externally, and Fisher, regarding the purported greater willingness of regional actors to both appreciate and foster such concerns – there is still a sustained analytical advocacy for seeking out and appreciating other potential partners in a wider stabilisation process. Regardless of the record thus far, with Berdal lamenting a 'recurring failure to acquire, let alone make use of, knowledge of local conditions and realities', he still advocates, as a corrective, a reprioritisation of attention away from the centre towards 'local, municipal and regional governance', which may have a greater appreciation of the peculiarities and priorities of local people.[57] In noting this centralising bias, Berdal and others are confirming part of what concerns Chandler in his chapter, although it is unclear whether a greater reprioritisation is either possible, or, from the centre's perspective, even desirable. In the case of both capacity building more generally, which may be viewed eventually as a challenge to central control rather than as a complement to it, and the wholesale transformation that would be required in the redirection of aid resources, given that one study of aid allocation noted that only 0.2 per cent of the cumulative Overseas Development Assistance (ODA) total sum went to local civil society organisations,[58] it seems more likely the international community will continue to direct its efforts primarily to keeping weak states under some kind of central control.

---

57 Berdal, *Building Peace after War*, 19.
58 Maria J Stephen, 'Adopting a Movement Mindset and addressing the Challenges of Fragility', *Fragility Study Group Policy Brief* 4 (2016), 4, available at https://www.usip.org/sites/default/files/Fragility-Report-Policy-Brief-Adopting-a-Movement-Mindset-to-Address-the-Challenge-of-Fragility.pdf.

There is a further balance to be struck in the process of reconstruction, relating to the shifting of responsibilities and decision making between external intervenors and internal central government over time. In what Dominik Zaum has referred to as 'the sovereignty paradox', whereby the external party or parties must continue 'compromising sovereignty in order to establish a sovereign state',[59] a delicate balancing act has to be maintained between permitting the local actors, at whatever level, to make decisions and be seen to make decisions – in order to increase their internal legitimacy as the source of key public goods and services – and maintaining a template for development that will help ensure continued external legitimacy and donor support. While legitimacy is in part based on what the state could be capable of doing, rather than solely the demonstrative effect of it actually acting, if such nascent sovereign capability is effectively emasculated by the strong centralising powers of a High Representative for Bosnia and Herzegovina, for example, or the perception the newly supported sovereign government is doing little more than reading from a pre-arranged Western script, then even the demonstrative impact of domestic effectiveness is ultimately lost. Given both the centrality and fragility of legitimacy, which 'can vary from village to village, from person to person over time'[60] and the likely continued existence of other forms of public goods delivery from pre-existing kinship groups, shadow state actors and other local alternatives, the intervening actors cannot afford, in the words of Immanuel Kant, to treat local actors as 'immature children, who cannot distinguish what is truly useful or harmful to themselves'.[61] Situations, such as that noted in Liberia, where a US-led SSR review so ignored local capacity that several central government security agencies did not even know it was taking place, let alone contributed to it, should be avoided,[62] with a preferred model of partnership being more in line with the Plan Colombia discussed by Rodríguez and Sánchez in their chapter, which has stood the test of time and political change thus far.

The issue of time as a measure of commitment – as well as a testimony to the complexities of state-building as a response to failings within a state – remains central to external intervenors' concerns. During 2019,

---

59 Dominik Zaum, *The Sovereignty Paradox: The Norms and Politics of International Statebuilding* (Oxford: Oxford University Press, 2007), 5.
60 Florian Weigand, 'Afghanistan's Taliban: Legitimate Jihadists or Coercive Extremists?', *Journal of Intervention and Statebuilding* 11:3 (2017), 374.
61 Kant cited Sorenson, 'After the Security Dilemma', 369.
62 Sarah Detzner, 'Modern Post-Conflict Security Sector Reform in Africa: Patterns of Success and Failure', *African Security Review* 26:2 (2017), 120.

the Trump administration was reportedly preparing to announce a final withdrawal from Afghanistan, after almost 20 years of military, political and economic engagement. Both the manner in which it was seeking to do so and the wider record of achievement engendered by such an extended international commitment to reconstitute the Afghan state provide telling lessons from the past and indicators for the future, in terms of willingness to commit in such a manner again. In the case of the former, as negotiations continued between US representatives and the Taliban, to create the conditions effectively to return the latter to some form of governance role, it was significant to note the exclusion of the elected and therefore legitimate government of Afghanistan from the process. Given what was argued above – and that the Afghan government can act as a spoiler itself, as it has done in the past, cancelling direct talks in April 2019 over disputes over who could attend, the size of each delegation and the role of female representation, and disputing the status of a Taliban 'embassy' in Qatar in 2013 – it does not aid the perception of internal legitimacy when talks are reported as having 'side-lined the Afghan government'.[63] Not only that, but a central concern – in line with Chandler's argument that 'problems there are always framed as problems for us' – relates to the conditions for a US departure and commitments to conditions that will not force it to consider returning, with leading US negotiator Zalmay Khalilzad quoted as saying that, while the US remains committed to the conditions of liberal democracy, including human rights and gender equality, 'they are Afghanistan's internal issues'.[64] This is in line with the Trump approach to the liberal world order broadly speaking and, perhaps perversely, injects a note of local ownership and conscious un-Western normative framing that much of the academic literature has focused on; although such concerns have been predicated more on not beginning the process with a pre-conceived Western notion of democratisation and liberalism in mind, rather than ceding such credentials as a means to finally depart.

In that sense, it was interesting the announcement of progress in talks with the Taliban was framed in some quarters as an 'abandonment' of Afghanistan, even after the level of commitment by the West, in terms of blood and treasure over two decades. It is difficult to see how the US could be accused of abandonment, given the sheer duration of its commitment

---

63 'Abandoning Afghanistan', *The Times*, 20 July 2019, 27.
64 Anthony Lloyd, 'We're back and ready to take power, declare Taliban', *The Times*, 20 July 2019, 39.

to reconstructing the state. On the other hand, the relatively limited outputs generated as a consequence of this sustained presence suggest that additional effort is insufficient as a mechanism for securing stability. After almost two decades, the statistics regarding Afghanistan do not make easy reading. At a cost of 2,372 American lives – not counting the figures for coalition partners, Afghan citizens, both civilian (believed to be approximately 32,000 in the last decade alone) and members of the Afghan National Army (ANA) and Police (ANP) – and a cost of approximately $1 trillion, the combined efforts of external intervenors, regional actors and local capacity had delivered a state where over 35 per cent of its territory remained under insurgent control in 2019.[65] Corruption at all levels remains rampant, with implications for the distribution of aid and other forms of international support, in a situation where, paradoxically, 'the countries that are most in need of reform also have the least capacity to implement it',[66] with concerns raised in the past about the very capacity that was lacking failing to equitably and effectively absorb international assistance.

There have been positives, with progress made in terms of human rights, gender equality, the provision of fundamental services, from education to electricity provision (with a February 2019 report assessing that the situation had improved from virtually zero in 2000 to 85 per cent coverage by 2018), and an increase in economic growth at the same time as a steady reduction in the reliance of the Afghan economy on external aid, which decreased from 206 per cent in 2006 to 59 per cent by 2015.[67] However, until mechanisms can be found to provide basic levels of security across the state as a whole, Afghanistan will remain a problematic case.

It may be – as with AMISOM in Somalia or efforts to stabilise South Sudan, a state so badly managed that it is not able to exploit 'its enormous national reserves', finding itself prey to 'a kleptocracy, a militarised, corrupt, neo-patrimonial system of governance' even before considering wider regional contexts[68] – that both academic and practitioner focus is drawn to the hardest cases to confront. Earlier intervention, while more difficult to justify – particularly if the consequences of state fragility or failure do not, per se, threaten the national security concerns of external actors – may prove

---

65 *The Times*, 'Abandoning Afghanistan'.
66 Ottaway and Mair, *States at Risk and Failed States*, 6.
67 Daniel Runde and Ambassador Earl Anthony Wayne, 'Finishing Strong: Seeking a proper exit from Afghanistan', *CSIS Briefing* (2019), 4, available at https://www.csis.org/analysis/finishing-strong-seeking-proper-exit-afghanistan.
68 Alex de Waal cited Johnson, *South Sudan: The Untold Story*, 41.

to be more effective. However, the international community, particularly the West, as it re-orients towards a more traditional foreign policy approach concerned with the return of great power competition, rogue states and the politics of nuclear disarmament, seems to have lost confidence and interest in a larger-scale model of international intervention, as can be seen in the hesitancy shown with regard to Syria and the desire by NATO to effectively sit on the sidelines in post-conflict Libya, brandishing the fig leaf of National Transitional Council (NTC) opposition to an extended external stabilisation mission as a means to salve collective consciences as Libya descended into chaos.[69] As with other aspects of the debate surrounding the mechanisms for intervention, there is a seeming binary switch being made – towards shorter term interventions and more limited processes of SSR, sometimes within the context of longer term commitments as in Afghanistan – on the basis less of their merits than unwillingness to consider a more extended approach. In his chapter, Pospisil notes the centrality of SSR within peace processes, although this cannot be because of its wider record of success; it may be as much because it is a lessened form of commitment on the part of external actors seeking to plug a gap rather than fix the fundamental framework. Elsewhere Ross, in noting a debate on SSR polarised between the view that it is 'a luxury, to be conducted in the spare time of military personnel not engaged in war or training' and the belief, born of wider frustration, that it is a 'magical device to resolve the irresolvable', has expressed justifiable concern that the end result is that 'neither embeds security cooperation within a fully realized strategy'.[70]

Matisek, in his chapter, also seeks to offer a little more balance: while indicating a series of cases where a process of military reform seems to have worked, he also notes the regular creation of what he calls 'Fabergé Egg armies' that are 'easily cracked', and he calls for the West to move away from its preferred model of civil-military relations to one that accepts a greater centrality for the military within the state as part of a 'strategic partnership between politicians, societal and military elites'. There is, sadly, plenty of evidence that, rather than providing a panacea, the West, in particular, has almost as many problems training an effective indigenous capacity capable of maintaining a legitimate monopoly of force. In the case of Afghanistan, having witnessed the lack of commitment to police and army reform

---

69 Brown, 'Kosovo and Libya'.
70 Tommy Ross, 'Leveraging Security Co-operation as Military Strategy', *Washington Quarterly* 39:3 (2016), 92.

initially,[71] the US acknowledged in 2017 that 'Afghan security forces cannot succeed – or even function – without sustained international assistance', with a 90 per cent illiteracy rate at the time one of many fundamental problems confronting international training and reform efforts.[72]

The academic literature is replete with examples of training efforts failing to translate into effective military force, most notably the Iraqi army's failure to resist IS in 2014, despite extensive US-led training and assistance since 2003, and it suggests that this process is not the quick fix it might have been assumed to be. Additionally, as Knowles and Watson critically assess in their chapter, a policy of increased drone strikes in order to create safe havens – a preferred method of intervention of both the Obama and Trump administrations – not only fails to even begin to address some of the deeper underlying concerns, focused as it is on symptoms and not the underlying condition, but is viewed by even those undertaking it as a means to limit and distance the external actor from effective engagement, with one of their interviewees dismissing the approach as tantamount to 'throw a few soldiers there and a few soldiers there'. Not only do Knowles and Watson argue – in line with much of the literature on drone strikes generally – that it may end up being counterproductive, generating continued instability where 'the use of remote warfare … may inadvertently yield exactly the kind of states within which terrorist networks thrive', but they posit the test as to whether a more limited approach actually 'prepares the ground for greater stability or whether it locks everyone into a never-ending cycle of violence'. As it cannot, in isolation or even as part of a wider approach, achieve the former, its limits become even more apparent.

Conclusion

The fascination with fragility continues unabated, perhaps encouraged by the continued ambiguity that exists regarding all areas of the subject, from definitional dilemmas to difficulties with responses, in all shapes and sizes. Academics are attracted by the ambiguity, even as it proves problematic for practitioners to even begin to learn the right lessons from each individual intervention. As has been demonstrated by the contributions to this book,

---

71 Coll, *Directorate S*, 129.
72 Colin D. Robinson, 'What explains the failure of US Army Reconstruction in Afghanistan?', *Defense and Security Analysis* 34:3 (2018), 251–57.

the barriers to achieving an effective and enduring intervention for the purposes of stabilising a fragile or failing state remain legion, ranging from the continued conceptual blinkers that limit external access to knowledge to the deleterious impact of a range of practical responses, from short-term drone strikes in lawless areas to creating more durable peace processes that anchor the longer term process of capacity building. Disputes continue over every aspect of the process, from the underlying drivers of weakness, which may shift from context to context, to the nature of the threat posed, with the literature at times repeating received wisdom more than challenging it, from the preferred vehicle for intervention – particularly problematic given that regional actors are seen as a better option by default rather than design – to the best mechanisms to seek in order to remedy weaknesses.

To (mis)appropriate Enoch Powell's pessimistic warning about the ultimate impact of all political careers, it seems that – even measured on their own terms – external interventions, of whatever kind, commitment and character, are ultimately destined to fail, with Collier noting that 50 per cent of peace processes fail within a decade.[73] Making binary switches – from top-down to bottom-up, international to regional, longer to short term – on the basis of the failings of previous experiences is not a recipe for success, as many of the underlying problems from more extensive internationally led approaches feature just as prominently in more limited locally oriented versions. While the problems posed internationally may not be as great as the immediate post-9/11 narrative may have suggested – with a preferred focus on earlier signs of fragility, rather than the happily rarer absolutist version of state failure, and a more critical appreciation of the actual nature of the threat posed – the humanitarian cost felt in contested states such as Yemen and Venezuela is real. Unfortunately, the Responsibility to Protect actually inculcated and practised by the international community more widely is not. Although commitment waxes and wanes – and, as the situation in Afghanistan demonstrates, is not, in and of itself, sufficient to guarantee success – it has not disappeared completely. It is only by acknowledging those limitations, in terms of knowledge, implications, interventions and impact, that a more realistic approach to tackling the phenomena of state fragility and failure can take shape.

---

73 Paul Collier cited de Coning, 'Adaptive Peacekeeping', 312.

References

Barakat S. and A. Larson. 'Fragile States: A Donor Serving Concept? Issues with Interpretation of Fragile Statehood in Afghanistan', *Journal of Intervention and Statebuilding* 8:1, 2014.

Bellamy, A.J. 'The Responsibility to Protect: Added Value or Hot Air?', *Co-operation and Conflict* 48:3, 2013.

Berdal, M. *Building Peace after War*. London: Routledge 2009.

Boutellis, A. and L. Connolly. *The State of UN Peace Operations Reform: An Implementation Scorecard*. New York: International Peace Institute, 2016.

Brock, L. et al. *Fragile States: Violence and the Failure of Intervention*. Cambridge: Polity, 2012.

Brown, D. 'Kosovo and Libya: Lessons Learned for Limited Humanitarianism?', *Comparative Strategy* 38:5, 2019.

Bruton, B. 'In the Quicksands of Somalia: Where doing less helps', *Foreign Affairs* 88:6, 2009.

Burgess, S. 'A Lost Cause Recouped: Peace Enforcement and State Building in Somalia', *Contemporary Security Policy* 34:2, 2013.

Carboni, A. and J. Moody. 'Between the cracks: Actor Fragmentation and local conflict systems in the Libyan civil war', *Small Wars and Insurgencies* 29:3, 2018.

Chesterman, S. 'Leading from Behind: The R2P, the Obama Doctrine and Humanitarian Intervention after Libya', *Ethics and International Affairs* 25:3, 2011.

Coll, S. *Directorate S: The CIA and America's secret wars in Afghanistan and Pakistan 2001–2016*. London: Allen Lane, 2018.

Cooke J.G. and R. Downie. *Rethinking Engagement in Fragile States*. Washington DC: Center for Strategic and International Studies, 2015.

de Coning, C. 'Adaptive Peacebuilding', *International Affairs* 94:2, 2018.

Detzner, S. 'Modern Post–Conflict Security Sector Reform in Africa: Patterns of Success and Failure', *African Security Review* 26:2, 2017.

Evans, G. and R. Thakur. 'Humanitarian Intervention and the Responsibility to Protect', *International Security* 37:4, 2013.

Gabrielsen Jumbert, M. 'How Sudan's 'Rogue' label shaped US responses to the Darfur crisis: What's the problem and who's in charge?', *Third World Quarterly* 35:2, 2014)

Gallagher, A. 'The promise of Pillar II: Analysing International Assistance under the Responsibility to Protect', *International Affairs* 91:6, 2015.

George, J. 'State Failure and Transnational Terrorism: An Empirical Analysis', *Journal of Conflict Resolution* 62:3, 2018.

Goldstone, J. 'Pathways to State Failure', in Starr, H. (ed.), *Dealing with Failed States: Crossing Analytical Boundaries*. London: Routledge, 2009.

Grimm, S., N. Lemay-Hebert and O. Nay (eds). *The Political Invention of Fragile States: The Power of an Idea*. London: Routledge, 2015.

Halverson, D. *States of Disorder: Understanding State Failure and Intervention in the Periphery*. Farnham: Ashgate, 2013.

Hammond, L. 'Somalia Rising: Things are starting to change for the world's longest failed state', *Journal of East African Studies* 7:1, 2013.

Healy, S. 'Rebuilding State Fragility: Somalia's Neighbours: a help or a hindrance?', in P. Lewis and H. Wallace (eds), *Rethinking State Fragility*. London: British Academy, 2015.

Hehir, A. 'The Myth of the Failed State and the War on Terror: A Challenge to Conventional Wisdom', *Journal of Intervention and Statebuilding* 1:3, 2007.

Hehir, A. 'Kosovo 1999: The False Dawn of Humanitarian Intervention', *Comparative Strategy* 38:5, 2019.

Hendry, K. 'From Failed to Fragile: Renaming the Index', *Fragile State Index 2014*. Washington DC: Fund for Peace, 2014.

Holm, H.H. 'Failing Failed States: Who forgets the forgotten?', *Security Dialogue* 33:4, 2002.

Kaplan, S.D. *Fixing Fragile States: A New Paradigm for Development*. London: Praeger Security International, 2008.

Lewis, A. 'Violence in Yemen: Thinking about Violence in Fragile States beyond the confines of conflict and terrorism', *Stability: International Journal of Security and Development* 2:1, 2013.

Logan, J. and C. Preble. 'Failed States and Flawed Logic: The case against a standing Nation-Building Office', *Policy Analysis* 560, 2006.

Loubser, H.M. and H. Solomon. 'Responding to state failure in Somalia', *African Review* 6:1, 2014.

MacGinty, R. 'Against Stabilization', *Stability: International Journal of Security and Development* 1:1, 2012.

Mason, M.C. 'Nation-Building is an oxymoron', *Parameters* 46:1, 2016.

Menkhaus, K.J. 'Vicious Circles and the Security Development nexus in Somalia', *Conflict, Security and Development* 4:2, 2004.

Menkhaus, K.J. 'State Fragility as a Wicked Problem', *Prism* 1:2, 2010.

Mezran, K. and E. Miller. *Libya: From Intervention to Proxy War*. Washington DC: Atlantic Council, 2017.

Miller, J. and K. Krause. 'State Failure, State Collapse and State Reconstruction: Concepts, Lessons and Strategies', *Development and Change* 33:5, 2002.

Nuruzzaman, M. 'Revisiting the category of Fragile and Failed States in International Relations', *International Studies* 46:3, 2009.

Ottaway, M. and S. Mair. *States at Risk and Failed States*. Washington DC: Carnegie Endowment, 2004.

Patrick, S. 'The Unruled world: The case for good enough governance', *Foreign Affairs* 93:1, 2014.

Polese, A. and R. Hanau Santini. 'Limited Statehood and its Security Implications on the fragmentation of political order in the Middle East and North Africa', *Small Wars and Insurgencies* 29:3, 2018.

Robinson, C.D. 'What explains the failure of US Army Reconstruction in Afghanistan?', *Defense and Security Analysis* 34:3, 2018.

Ross, T. 'Leveraging Security Co-operation as Military Strategy', *Washington Quarterly* 39:3, 2016.

Rotberg, R.I. 'The New Nature of Nation-State Failure', *Washington Quarterly* 25:3, 2002.

Rotberg, R.I. (ed.). *When States Fail: Causes and Consequences*. Princeton NJ: Princeton University Press, 2004.

Sorenson, G. 'After the Security Dilemma: The Challenges of Insecurity in Weak States and the Dilemma of Liberal Values', *Security Dialogue* 38:3, 2007.

Stephen, M.J. 'Adopting a Movement Mindset and addressing the Challenges of Fragility', *Fragility Study Group Policy Brief 4*, 2016.

Thakur, R. 'The Responsibility to Protect at 15', *International Affairs* 92:2, 2016.

Weigand, F. 'Afghanistan's Taliban: Legitimate Jihadists or Coercive Extremists?', *Journal of Intervention and Statebuilding* 11:3, 2017.

Weiss, L. *The Myth of the Powerless State*. New York: Cornell University Press, 1998.

Zaum, D. *The Sovereignty Paradox: The norms and politics of International Statebuilding*. Oxford: Oxford University Press, 2007.

INDEX

www.ingramcontent.com/pod-product-compliance
Lightning Source LLC
Chambersburg PA
CBHW060318030426
42336CB00011B/1113